HOW TO DISAPPEAR AND LIVE OFF THE GRID

HOW TO DISAPPEAR AND LIVE OFF THE GRID
A CIA INSIDER'S GUIDE

JOHN C. KIRIAKOU

Skyhorse Publishing

Skyhorse Publishing books may be purchased in bulk at special discounts for sales promotion, corporate gifts, fund-raising, or educational purposes. Special editions can also be created to specifications. For details, contact the Special Sales Department, Skyhorse Publishing, 307 West 36th Street, 11th Floor, New York, NY 10018 or info@skyhorsepublishing.com.

Visit our website at www.skyhorsepublishing.com.

10 9 8 7 6 5 4 3 2 1

Library of Congress Cataloging-in-Publication Data is available on file.

ISBN: 978-1-5107-5612-0
eBook ISBN: 978-1-5107-5613-7

Cover design by Kai Texel

Printed in the United States of America

For Kate

Contents

Author's Note

All statements of fact, opinion, or analysis expressed are those of the author and do not reflect the official positions or views of the US government. Nothing in the contents should be construed as asserting or implying US government authentication of information or endorsement of the author's views.

Introduction

History is filled with accounts of people who have disappeared. From the early days of American history there were mysterious tales such as that of the Roanoke colony which disappeared while it waited for supplies from England. To this day, historians speculate about what happened to the group of just over one hundred English settlers. Sometimes the reasons for disappearances are obvious such as that of Amelia Earhart, who vanished while on a solo airplane flight around the world. Other people while hiking in the forest become lost and are never seen again. And then there are the people who disappear not as a result of misadventure but because they need to hide.

An examination of people who have managed to escape shows that the best way not to be found is to isolate yourself from your world. This presents enormous risks as technology is advancing so quickly. The small place that you have chosen may be found from a satellite image, from a helicopter, or even a drone. Then there are the curious hikers who may get lost from their track as they are searching for a great adventure and find themselves in the middle of your new home. Unless you're Eric Rudolph, the domestic terrorist who bombed the 1996 Atlanta Olympics and then took to the woods before being found in 2003, you likely won't want to just live out in the woods somewhere.

Our culture often glamorizes those who have taken flight whether from law enforcement or criminal gangs. In Mafia films it's often called "going on the lam" or "going to the mattresses." Many people still remember the Simon and Garfunkel hit song "Somewhere They Can't Find Me" which romanticizes the story of a young criminal who bids farewell to his girlfriend by telling her that his "life seems unreal" as he is forced to fly down the highway. Contemporary fiction, for the purposes of entertainment, often examines the issue of disappearances. One of the best is the author Thomas Perry who in the novel *Vanishing Act* introduced a character named Jane Whitefield who is a Native American guide who works to helps desperate people escape from enemies who want them dead. As a member of a clan of the Seneca tribe, she has a supportive network of people who can provide a safe haven while she provides her clients with new identities that are backed up by authentic paperwork. With her special skills, she can create deceptive trails that will defeat those who pursue her clients. Her motives are pure and she does not work for money, while her clients are not criminals, but honest people who can no longer survive with their real identities. Though Jane Whitfield is a fictional character, there are certainly people who, for a price, can help someone disappear.

Several decades ago, there was a popular television series entitled *The Fugitive*. It was the story of a doctor, Richard Kimble, who was wrongly convicted for the murder of his wife. The starting point for the program is when he escapes custody and has to hide from the police while searching for the one-armed man who actually killed his wife. In 1993, the story was made into a very popular movie and, as in the television series, viewers were encouraged to cheer for the fugitive.

People disappear for a variety of reasons. In the world of intelligence operations, agents often need to disappear when they have accomplished their mission. And then there are the unhappy spouses who want to start a new life unencumbered by the consequences of a failed union. Some disappearances are understandable and fully justified while others may involve a flight from justice if you are being sought for a criminal action. Prospects for eluding the police are reasonably promising. A key factor is the severity of the crime for which you are being sought. Something like aggravated assault will not place a person on the FBI Most Wanted list, whereas kidnapping, murder, or bank robbery will ensure your prominence. Because law enforcement agencies have limited resources, many cases simply disappear through official inattention. In an area like Las Vegas, each year more than 400,000 arrest warrants are issued, a brief search may be undertaken in the city, and then the warrants are forgotten. For fugitives whose crimes are lesser and have not generated public outrage, the most likely way in which they will be caught is in a routine traffic stop or some other mundane event. Otherwise, for most people on the run, freedom begins at the county line.

Our freedom may begin at the Canadian border since it is relatively easy to leave the United States if you are the object of a police search. There is no border coverage at all in remote areas of Maine, Minnesota, Montana, and North Dakota, so a fugitive can simply walk into Canada. Because of a relatively open southern border, it is also possible to make it into Mexico. However, both Canadian and Mexican authorities are alert to the presence of such fugitives and usually will quickly apprehend and imprison them.

American authorities have expressed concern about the problem of criminal suspects who have found havens in Canada and Mexico. US Senator Dick Durbin (D-IL) has lobbied for a strengthening of the US-Mexico extradition treaty because it omits many serious crimes. The US Justice Department has established special initiatives to apprehend border-crossing fugitives.

Perhaps most frequently discussed is the person who needs to disappear because he is being sought by the police for some criminal act he committed. Depending on the severity of the charges he faces, this person is highly motivated and is especially aware of the desperation of his situation. A failure of his flight plan can mean confinement in a correctional institution for many years or perhaps even execution. People in this category, as fugitives from justice, stand out on statistical summaries because they are reported. From the internet, you can get an approximate number for this category on a given date. The unhappy spouses or the people whose lives have become too difficult are not necessarily counted so the internet cannot tell us how large this community may be. What we do know is that, according to the National Missing and Unidentified Persons database, more than 600,000 people of all ages go missing every year in the United States and well over four thousand unidentified bodies are found each year. Some cases, like that of Elizabeth Smart, get national attention but most cases never get much more than brief local attention. Most of the approximately 600,000 cases are resolved because the people returned or were determined to have died of natural causes. In many cases, the alleged disappearance was nothing more than a misunderstanding. When all of this is sorted out it means that each year around two thousand people actually disappear.

There are numerous cases in which adults just decide to leave friends, family, and other connections behind in order to start over. This is not illegal, and someone's absence will not likely attract police attention. Police involvement will occur if it is a child who has gone missing or a dementia patient who is likely to die of exposure if not located. In recent years there has been a decline in the reported number of missing persons in the United States. One reason for this is the improvement in our ability to communicate through programs such as the Amber Alert and the installation of closed circuit cameras in most large cities.

Kidnapping is an important factor in disappearances each year.

This is a significant issue with missing children, as 1,435 children are kidnapped each year. Most of these cases are family members kidnapping a child. Because the perpetrator is a family member, the authorities will report less information. More serious are kidnappings by non-family members. Each year there are more than two hundred such abductions and these are cases that get more attention. Kidnappings of adults are more common and each year FBI statistics indicate that approximately sixty-five thousand adults are apparently kidnapped. Since a kidnapping will involve the FBI, a fake kidnapping is never a good cover for a person who simply wants to disappear.

The plight of the abused spouse is a frequent situation requiring the spouse to disappear. Abused women continually face threats that require them to acquire a new identity or to find a new residence. Their pursuers are not likely to be law enforcement officials who would follow the law and behave in a predictable manner. It is more likely to be a vengeful former partner who will be not subjected to legal limitations in their efforts to track down the fugitive spouse or girlfriend. They will do the searching themselves or, if they have the resources, they will enlist the aid of a professional. In either case, the pursuit will be much more brutal.

Closely related to the abused spouse category are victims of stalkers. According to the Rape, Abuse, & Incest National Network, each year more than three million people are victims of stalkers. Approximately 250,000 are eventually raped or killed by their stalker. While most victims of stalkers do not have to go into hiding, many do and must disappear from their home communities.

In more recent circumstances many people feel the need to go underground for political reasons. The acrimonious political environment in the United States, for example, has led to a situation in which officials and politicians label their adversaries as either traitors or even terrorists. While it is difficult to predict if this bitter political divide will persist, many people are being denied internet platforms because authorities suggested their rhetoric "might" lead some people to become violent. Others also face severe economic deprivations as their victorious political adversaries use their economic power to suppress their vanquished foes. As of now, some people have discussed flight as a future option. It may well be that our hostile political, social, and cultural environment will create an entirely new class of political refugees. As will be discussed later, political fugitives became a common phenomenon during the years of anti-Vietnam war protests when dissidents engaged in bombing, hijackings, and armored car robberies. The perpetrators of these actions typically fled, assumed new identities, or even went overseas to Canada, Cuba, or Algeria.

Even for those not in full flight from some sort of political repression, the ease with which the government can monitor all of our communications is a growing concern. Thousands if not millions of people have abandoned many of the established and most powerful social platforms. The economic losses suffered by Facebook, Twitter, and Instagram are considerable. Critics have turned to other platforms in which there is less fear of censorship. Among the most successful of these alternatives was Parler. By coordinating their repressive responses, the social media giants, including Amazon, were able to effectively destroy Parler within just a few days. When WhatsApp released a new privacy policy that gave Facebook access to WhatsApp data, many people turned to Telegram and Signal as more secure and less censored platforms. An advantage of Signal is that only your phone number is linked to you. Telegram will link your name, your number, contact list, and user ID. Telegram also requires users to opt into its "secret chat" if you want your messages to be decryptable only by the recipient. Otherwise, you will have less privacy than on Signal. By contrast, WhatsApp collects a much wider range of data on users, including their locations and interests meaning that you have minimal privacy. Signal is funded by the non-profit Signal Foundation and Telegram is funded by the social media billionaire Pavel Durov.

The continuing dramatic growth of surveillance system undermines the privacy of most individuals, whether they are criminals or law-abiding citizens. The surveillance grid includes highly sophisticated security cameras with facial recognition capabilities. Behavior that might look suspicious can generate a computer search that within seconds can produce an exhaustive summary of every aspect of your life. Tech empires such as Microsoft and Motorola can compile an enormous volume of information including everything from social media to police records. Working with other corporate entities, they have created *fusion systems* that can link surveillance grids not just in the United States but in other nations around the world. Fusion systems can bring together data from an unimaginable number of sources. They operate on a "correlation engine" that can identify behavioral patterns that will identify the likely suspects for almost any criminal activity. The engine can do much more than search for criminals. It can also analyze the political, social, and economic activities of private citizens in such a way as to generate a profile that will allow authorities to place them within carefully defined categories based on their expressed political attitudes. There is also a joint European surveillance system that can monitor anything regarded as hate speech or nationalistic attitudes.

It is important to note that if your privacy requirements can be met by platforms like Telegram and Signal, you are not in the difficult situation of someone who needs to actually disappear. These devices are useful for those who have a need for privacy but have not yet been reduced to using things that are drawn from the world of spycraft. In short, they do not need to disappear but only to keep a low profile that would help shield them from the sort of intense scrutiny that might prompt them to disappear. As long as their activities do not raise suspicion, discretion will help them avoid becoming a full-time refugee. Keeping a low profile provides protection from police observation but can help you avoid the interest of criminals who hope to prey on you.

Other people seek to flee from their current environment for the most mundane and excusable of reasons. With the advent of widespread and severe COVID restrictions in 2020, many New

York residents took flight to escape these difficult conditions. Often that flight was an inward examination of simple things such as the organization of their apartments. By contrast, many urban dwellers hope to find a better location, not wishing to actually disappear but rather to find what would be regarded as "greener pastures." The architects of their flight were not the purveyors of new identities, but simply architects who specialized in a renewed, innovative use of resources. Their pursuer was not someone intent on imprisoning or killing them but simply the stressful conditions of their city in the COVID era. As a result, people like this do not fall into the category of "disappeared" and are not being pursued by law enforcement authorities.

Such people will cite the desire to limit their carbon footprint or to live closer to nature. Their motivations and methods are not the products of desperation but the lessons of their reinvention of themselves are relevant to the fugitive being sought by other people, rather than a pandemic. Individuals who are mostly concerned about personal happiness in the most trivial sense will still want access to some modern conveniences. New Zealand residents recently boasted about founding an entire island that is off-grid and is home to a mere eight hundred residents. They are required to generate their own power and to develop sewage and water systems. Their goal is comfort while that of a fugitive is survival. Nonetheless, these are issues with which the fugitive must contend so the knowledge is valuable.

Whether you are a concerned environmentalist, an emergency prepper, or a potential fugitive, you have to undertake planning if you are serious about surviving. This may involve identification of what will become your "bug out" location. By locating a place to which you can flee, you will be able to stock the supplies needed for long-term survival off the grid. While planning ahead is prudent, it is important to note that there are some important disadvantages in this scenario. If your flight is prompted by a widespread societal breakdown, there is a strong possibility that criminal gangs who have learned that you have made careful plans will follow you to that hidden location. If you fail to detect the presence of such gangs, you will be at their mercy. You will certainly lose your supplies and possibly your life.

If you are operating during a societal breakdown, another possibility is that roads to your escape will be blocked. This could happen because of a police lockdown or, more likely, the activities of gangs planning to attack people who are fleeing from a city. Because police forces will be severely stretched during such a time, travelers who encounter gang blockades on the highway will not be able to count on police protection. If your flight is prompted by a natural disaster, this is another instance in which you may not be able to get to your hidden sanctuary.

Regardless of the reasons for your disappearance, the planning process is daunting. You have to make basic decisions about whether you will hide in plain sight or escape into the wilderness. If hiding in plain sight in the United States, there are some crucial considerations in your selection process. States with large, diverse populations are more promising. The existence of a well-developed system of mass transit is something else that helps the fugitive. New Jersey, Ohio, and Florida offer these advantages as well as already having large populations of fugitive felons.

It is also important that you consider international flight. This is especially difficult but does offer the prospect of living in reasonable comfort in another part of the world. If an international flight is what you prefer, the next step is to identify regions of specific countries. That means you must understand conditions in those nations so you can pick one that is likely to be more welcoming. The range of possibilities has varied greatly over time. In the immediate post-World War II era, few places were more welcoming than Tangiers, Morocco. When order was restored, Marseilles, France, emerged as an attractive prospect since it was under the control of French organized crime groups. In the 1970s, Thailand became a highly favored destination with Paraguay gaining prominence in the next decade. After 2000, Lagos, Nigeria, gained a reputation that was attractive to many fugitives, although the Nigerian criminal environment eventually began to discourage all but the most vicious of fugitives. In general, fugitives were drawn to places in which the concept of extradition was very important and the local environment was suitably chaotic. With this, Malta, Yugoslavia, and the divided Cyprus gained in popularity.

Because of its proximity to the United States, many people think of Mexico as a suitable place of refuge. Because the country is associated with considerable lawless behavior, there is an assumption that an American could easily live there as a fugitive. This is actually far from the truth. Extradition laws in Mexico are similar to those in any state in the US and Mexican authorities are not sympathetic to American criminals who select their country as a safe haven.

Speculation about countries that might be suitable homes for fugitives usually focused on three issues. The first is population density of the country. If you hope to avoid human interaction, you would want to avoid nations with large, congested populations. A second issue is distance from the United States. A place like Canada, which is next door to the US, is not as attractive as a country in Africa where you are less likely to encounter fellow Americans. The final issue relates to the existence of an extradition treaty with the United States. Mexico has a well-defined extradition treaty with the US and should probably be dismissed as a possible haven for felons who are on the run. There are approximately seventy-five nations that do not have extradition treaties with the United States. Most prominent among this group are several former republics of the USSR. Russia is best known in this respect since providing a safe haven for NSA whistleblower Edward Snowden. Other post-Soviet states with no extradition treaties are Moldova, Armenia, Belarus, Kazakhstan, Uzbekistan, and Ukraine. Not surprisingly, the Peoples Republic of China and Libya are also on this list.

The countries which offer the best prospects for fugitives are not in Europe but, for the most part, in the Third World. That does not mean that a person could not live in relative comfort in most of the locations. Kuwait and Lebanon are prosperous and enjoy modern amenities such as good health care as well as beautiful scenery. A fugitive with a reasonable amount of money could have a comfortable life in Kuwait and Lebanon. Bangladesh features a pleasant mix of Indian and Asian culture. Burundi and Rwanda offer mountains and lakes. Rwanda, which was known for devastating atrocities in ethnic fighting in the 1990s has enjoyed a remarkable recovery and is now an attractive locale for the discerning fugitive. The same can be said for Vietnam, which has recovered from decades of conflict and has become an attractive tourist destination, as well as good place for fugitives. The South Asian Maldives islands are most frequently cited as the most comfortable get away for a person who is on the run from US authorities. Even non-fugitives are attracted to what is regarded as one of the world's most beautiful locations for either a short tourist visit or a life-time safe haven.

If you are going to choose an overseas location, there should be an additional step in your planning. Before moving to the country you select, take the time to actually visit it as a tourist. In making the journey, do not leave indications of your interest. Visits to their embassy or lengthy searches on the internet will serve as clues if someone sets out to locate you. In traveling to this country, do not take a direct route but make it appear that in your apparent vacation, you were interested in several places. You might even be able to find a package tour that includes this country. Remember, secrecy is your best friend.

Dichotomy of Pursuers

ONE OF THE VERY FIRST THINGS to consider when thinking about going underground is to identify the nature and outlook of those who have created your need to disappear. There is a big difference between running from law enforcement authorities as opposed to dodging an angry spouse. The latter might be able to enlist the services of a private detective but would have to personally cover those expenses, which are considerable. Law enforcement authorities are paid by taxes and have fewer logistical constraints in their efforts. A chase that is measured in years is easily managed as long as the superiors decide it is justified.

The most likely worst case for the fugitive is to be pursued by law enforcement with its abundance of resources. History offers numerous instances of the intrepid police officers who never gave up in their search for wrongdoers who might have been wanted for heinous crimes such as rape, kidnapping, robbery, or murder. While these criminals may have fled overseas and attempted to live with new identities, relentless detectives did not give up. Their stories often become the stuff of legend, literature, or histories of law enforcement agencies.

Television programs such as *America's Most Wanted,* which ran for twenty-three years, illustrated the ability of law enforcement to enlist the media and the public in their searches. The show was the longest-running program in the history of the FOX television network at the time. Before the advent of such modern facilities for identifying a fugitive, there was little more than the wall in the post office reserved for wanted posters. If the target was especially threatening, posters might appear in a variety of storefronts but they still reached a modest number of citizens. With programs such as *America's Most Wanted*, civic minded citizens, or simply bored people with time on their hands could join in the hunt.

David James Roberts enjoys the dubious distinction of being the first person profiled on *America's Most Wanted* who was apprehended as a result of the show. Roberts had been convicted of murder in 1975 and for kidnapping a woman while out on bail in 1977. He escaped in 1987 while being transported to another prison and was identified by callers to the show in 1988 who knew him as Bob Lord and who was working as the director of a homeless shelter in New York. Known as a gentle and compassionate man who was devoting his life to helping the homeless, his exposure came as a shock to those who knew him.

As a result of the popularity of shows involving the public in the pursuit of criminals, there were numerous spinoffs and copycat shows. FOX created the short-lived *America's Most Wanted: Final Justice* and other networks created programs like the ever popular *Unsolved Mysteries* that featured dramatic reenactments as well as updates regarding outstanding cases. The response to these shows was enthusiastic and, as noted above, their tip lines often received useful information from private citizens who believed they knew the identity of some of the individuals profiled in the program.

Other private pursuers might be bounty hunters who are motivated by profit but not limited by some of the legal niceties that inhibit professional lawmen. The bounty hunter—who is also known as a bail agency enforcer, fugitive recovery agent, or a bail enforcement agent—earns his money by capturing fugitives for a commission or a bounty. He is employed by the bail bondsman who has created a civil contract between himself and a defendant. In this capacity, he was an independent contractor, not a police officer, and may face liabilities from which a policeman is immune. In short, he can be sued, as many bounty hunters are each year. On the other hand, the bounty hunter can enter the home of a fugitive without having a search warrant or probable cause. If he enters the wrong home, he will face a lawsuit and, in some circumstances, may be subject to arrest. Entering the wrong residence is a common failure of the bail agency enforcer. Bounty hunters are represented by a professional association known as the National Association of Fugitive Recovery Agents.

Over the years, it has become routine for law enforcement to employ the news media in their efforts to apprehend fugitives. Most television stations now feature a Crime Solvers segment inviting citizens to call if they think they know anything about someone who is wanted for their criminal acts. The typical Crime Solvers program will provide names and photographs of alleged offenders as well as details of the crimes for which they are being sought. In Britain, police will often post signs at the sight of a criminal event or even an accident asking pedestrians to report if they saw anything at that particular location.

In modern times, the practice of using a bounty hunter is illegal in most nations and its acceptance varies from state to state in the United States. Commercial bail bonds are illegal in Illinois, Oregon, Kentucky and Wisconsin. A prospective fugitive needs to be aware of state laws regarding this practice as it will have a significant impact on the character of the pursuit they will face. When making an arrest, the bounty hunter may wear protective garments, display badges, and wear jackets emblazoned with BAIL ENFORCEMENT AGENT. Some will employ firearms while others limit themselves to non-lethal weapons such as pepper spray.

The training and authority of a bounty hunter will vary according to the state in which he is operating. In some states, neither training nor a license is required for them to pursue fugitives. In others, they must have both. If the fugitive knows something about the bounty hunter who is following him, as well as the requirements of the state in which they are located, the fugitive can travel to another state in which the bounty hunter has less or no authority and, in so doing, avoid arrest. In some states, the bounty hunter is not supposed to wear colors normally worn by local law enforcement officers or drive vehicles the same color as are used by local police.

The saga of the longtime fugitive is an old one and the recorded history of criminal behavior provides numerous examples. One of the more carefully documented is that of an English thief, Jack Sheppard, who had been imprisoned in 1724 after being arrested and imprisoned five times. While he may have been easy to catch, it was hard for law enforcement officials to keep him in custody. After escaping from prison four times, he became a notorious figure who enjoyed widespread popularity among England's poor citizens, who saw him as some sort of Robin Hood figure. His fugitive existence was maintained by his ability to create disguises such as that of a beggar. Eventually, he returned to London where he broke into a pawnbroker's shop and stole a silk suit, rings, and watches. With this apparel, he created another disguise, that of a dandy, who enjoyed the company of women while drinking to excess. It was his drunkenness that led to his capture, conviction, and execution.

William "Boss" Tweed, an American politician who was the leader of the Democratic Party political machine known as Tammany Hall, that enjoyed prominence and power in New York in the nineteenth century, is another important example of how the forces of state authority can foil an escape plan. As a result of his blatant corruption, Tweed was convicted and imprisoned in 1871. Because of his political influence, he was allowed to have home visits and used them as an opportunity to assume a new identity, that of a seaman, and he escaped to Spain. The extensive resources of the US government enabled authorities to learn of his plan and arrange for his apprehension once he got to Spain. As an eminently recognizable person, he was caught and returned to the United States where he remained in prison until his death two years later.

Another politician who was facing questions about corruption was New York City mayor Jimmy Walker. Walker, assisted by New York governor Al Smith, who liked his personal style and charisma, was elected mayor in 1925. The hallmark of his administration was blatant corruption but he did bring about popular changes, such as Sunday baseball. When the Great Depression hit, attitudes toward the flamboyant and corrupt Walker changed. In 1932, as criticism and suspicion increased, the national Democratic party felt that he might hurt Franklin D. Roosevelt's electoral prospects. He was denounced by the Archbishop of New York for his sinful ways and for his personal corruption by a commission led by Justice Samuel Seabury. Facing all of this, Walker decided to resign. Accompanied by Ziegfeld dancer Betty Compton, Walker took large fortune and went to the French Riviera where he demonstrated the fine art of hiding in plain sight. His change of scenery was not permanent and he was not pursued by either law enforcement or the curious journalists. Once interest in his case subsided, he was able to return to New York without facing criminal prosecution.

A man who is often regarded as America's first serial killer, H. H. Holmes, was the target of a long interstate pursuit when he emerged as a suspect in a series of murders during the Chicago World's Fair. Holmes was a skilled confidence man and operated a hotel near the fair site. He used it to lure women whom he would then brutally kill.

In 1894, under suspicion because of reports about his suspected behavior, he disappeared, and a search of the hotel revealed his ingenious methods for entrapping young women. The police demonstrated a great commitment in their search for Holmes. After a thorough investigation, he was traced to Boston, arrested, and eventually sentenced to death.

Another well-known story of people attempting to disappear in order to escape retribution for their criminal deeds is that of the Western outlaws known as Butch Cassidy and the Sundance Kid. They were the targets of a large number of pursuers over a period of years and over two continents. Born as Robert LeRoy Parker and Harry Longabaugh, the two men robbed banks and trains over a period of almost two decades in the latter nineteenth century. The two, along with occasional confederates, were pursued by both Pinkerton detectives and numerous deputies working for sheriffs from areas in which they committed their criminal acts. As the Pinkerton agents closed in on the criminals, they fled to New York City and then to South America in 1901. For a period of four years, they owned and operated a ranch in Argentina under assumed identities. With the decline of their legitimate activities, they once again turned to their outlaw trade and robbed banks, trains, and mine payrolls. The most common account maintains that the Pinkerton agents were assisted by Bolivian soldiers who engaged the two in a gunfight in 1909 that ended with the deaths of both Butch Cassidy and the Sundance Kid. Their romantic story was memorialized in a very popular movie, *Butch Cassidy and the Sundance Kid*, with their roles being played by two of America's most popular actors. The account of their flight would not be complete without noting other accounts such as their having died during a bank robbery in Uruguay in 1911 or even having traveled back to the United States and eventually assuming a new and respectable identity and writing an account of their criminal exploits entitled *The Bandit Invincible* in the 1930s.

The eventual apprehension of Bonnie and Clyde, who lived on the run for years, illustrates the strength of law enforcement. It was the Texas Rangers who made catching Bonnie and Clyde a priority and were able to enlist others who were determined that the criminal duo that had ridiculed law enforcement would not escape. Although the two had become folk heroes in a Depression-era America, authorities were able to pressure known associates of Bonnie and Clyde into giving them a least some information about the pair's movements. In 1934, six police officers from Texas and Louisiana set up an ambush on a country road in Bienville Parish, Louisiana. Most of these officers had followed Bonnie and Clyde from state to state during their flight from the law. This saga was also the subject of a popular movie which created a sympathetic image of the two criminals.

There are many stories about the skillful escapes of Nazi war criminals who fled from Germany and lived under assumed identities for decades. The key to their success was their careful planning and the organizations that facilitated their escapes, which generally involved flight to another continent. The German communities throughout Latin America played an important role in supporting their new identities. Among the most notorious was Josef Mengele, the "Angel of Death," who was an SS officer and a physician who conducted brutal experiments on inmates at the Auschwitz Concentration Camp as Germany fell and the camps were being liberated, Mengele was transferred to a different camp and went into hiding. In 1949, using an assumed name, he traveled to South America. Although numerous governmental agencies and police organizations, as well as the Israeli Mossad, sought Mengele, he successfully eluded them all. His death in 1979 did not come at the hands of any of these Nazi hunters but was the result of a stroke he suffered while swimming. Even in death, he was not exposed and it was not until the discovery of his remains in 1985 and DNA testing in 1992 that authorities were able to confirm that the Angel of Death had died.

Another Nazi war criminal who eluded authorities for several decades was Adolf Eichmann. While he did not enjoy the success of Josef Mengele, he was able to lead a secret life for fifteen years.

Eichmann was an Austrian Nazi who fled to Germany and became active in the Nazi movement. When Germany annexed Austria in March 1938, Eichmann led a raid on the Jewish Cultural Community office in Vienna and organized a Central Office for Jewish Emigration. In 1938, in recognition of his success he was allowed to create a program for the expulsion of Jews from the Protectorate of Bohemia and Moravia. This work ensured that he eventually became responsible for deportation proceedings in the Third Reich.

In short order, he took over responsibility for the deportation of Jews from all over Europe to the concentration camps in which they were killed. Because of his prominence, he was a participant in the Wannsee Conference and helped develop plans for the so-called "Final Solution." He also participated in much of this program as its primary organizer.

At the end of the war, Eichmann was held by American authorities, but managed to escape in 1946. By using the services of the Nazi "rat line," he made his way to Argentina, where he lived under a number of aliases. There were several Israeli Mossad agents who devoted years of their careers to finding Eichmann. He eluded capture until 1960, when Mossad agents located him in Argentina, kidnapped him, flew him to Israel. He was placed on trial and convicted for his crimes. In December 1961 Eichmann was sentenced to death and in May 1962, he was hanged.

Frank Freshwaters was an American criminal who eluded the authorities for fifty-six years. He had been convicted of a hit and run killing in 1957 and sent to an Ohio prison where he impressed prison authorities who gave him the honor of being a trustee. He was placed at a less restrictive prison work camp from which he escaped in 1959. During his decades as a fugitive he lived in three states under an assumed name, William Cox, and worked as a truck driver. He was never captured, but made arrangements to turn himself in to prison authorities. Because of an outpouring of public support organized by his attorney, the parole board voted to release him and place him on parole for five years.

As the Mengele and Freshwaters cases demonstrated, sometimes law enforcement, with its abundant resources, may be unable to apprehend the fugitive. The long and mysterious case of D. B. Cooper, who in 1971 skyjacked a Boeing 727 and held crew members hostage, is one of the most remarkable instances of a long-term search for a fugitive. Cooper parachuted from the airplane carrying $200,000 in ransom money and went missing in the thick Oregon forest below. The FBI finally admitted defeat, but private citizens conducted a lengthy search before apparently resolving this case. It was a veteran television newsman, Tom Colbert, who hoped to produce a documentary and brought this long investigation to what might be a resolution in 2013. While "resolved" may be a term that implies a greater degree of success, it is apparent that one investigator eventually—fifty-five years later—identified the likely perpetrator, a US Army veteran named Robert W. Rackstraw. This conclusion although did not result in a prosecution, a documentary, or even a confession, however. The story of D. B. Cooper has entered our popular culture and most people seem to accept his case as evidence of a successful flight from prosecution.

The case of the American criminal Donald Richard Bussmeyer, who had the distinction of being on the FBI's Ten Most Wanted list in 1967, illustrates the efficacy of the combination of law enforcement and an alert public. Bussmeyer was a career criminal who was convicted of auto theft, attempted burglary, assault with intent to kill, and armed robbery. In March 1967, he helped rob a Los Angeles bank and immediately left town with his two accomplices. Due to broad media coverage and publicity, the FBI enjoyed the benefit of widespread media coverage and within two months had captured Bussmeyer. The speed of his apprehension was, at least in part, due to the fact that he fled without having the necessary support in place.

One of the most notorious American fugitives was the serial killer and sociopath Ted Bundy, who was also a kidnapper, a rapist, and a burglar. His successful flight from numerous law enforcement jurisdictions is a tribute to the value of thorough planning. While he was imprisoned in 1975 on an aggravated kidnapping charge, he was suspected of being guilty of much more serious crimes in several states. Obviously realizing he was in a difficult situation, Bundy managed two escapes. The first took place in 1977 when he was transported to the Pitkin County Courthouse in Aspen, Colorado, for a preliminary hearing. Since he was acting as his own attorney, the judge had his handcuffs and leg shackles removed. During a recess, he asked to use the courthouse library from which he escaped through a window. He calmly walked past several checkpoints and was free for six days until police captured him during a routine traffic stop. Back in jail, he immediately began planning for another escape, which he effected at Christmas time while jail guards were less attentive. He got into a crawl space and came out in the unoccupied apartment of the chief jailer, stole his clothing, and walked to freedom. Afterward, he made his way to Florida where he committed three murders, as well as a variety of other crimes. He was eventually recaptured in 1978 in Florida where he had been living under an assumed name. He received three death sentences in a Florida court and was executed at Florida State Prison in 1989. Before his execution, he confessed to thirty murders in seven states over a period of four years.

While a private citizen may be involved in the pursuit of a fugitive, such an individual usually does not have an abundance of resources. However, a highly motivated person may be able to marshal resources equal to those of the police. The best example of this is seen in the work of John Edward Walsh Jr. whose son Adam was murdered in 1981. In order to find justice for his son, Walsh became a criminal investigator, as well as a victim rights advocate, and created the television program *America's Most Wanted*, which helped organize what were essentially private searches for fugitives. Through his work, his son's killer, a deceased serial killer named Ottis Toole, was finally identified in 2008. Even though Toole escaped prosecution for the murder of Walsh's son, he was posthumously exposed as the killer.

Over the last few decades, law enforcement organizations have sharpened their skills of pursuit and have been able to overwhelm the best efforts of many fugitives. In the 1970s, the FBI joined with the New York City Police Anti-Terrorist Task Force in an effort to pursue some of the most sinister and elusive terrorist groups in American history. This innovation led to the more recent creation of the Joint Terrorism Task Forces which have enabled law enforcement to further increase its efficacy.

It is difficult to say who might have been the most successful fugitive in recorded history. After all, the greatest tribute to their success must be that they were never caught. While that may be the result of their criminal proficiency, it might well be because in their flight they fell in a river never to be found. As noted above, Frank Freshwaters spent fifty-six years on the run. Known as Britain's most wanted man, Ronald Arthur "Ronnie" Biggs was a British criminal who took part in the Great Train Robbery of 1963. Biggs was not a sophisticated master criminal. He had never been out of England until his escape but he was able to travel through France, Australia, Argentina, Panama, Bolivia, and Brazil over a period of thirty-six years. In his autobiography, *Odd Man Out: The Last Straw*, Biggs discussed the difficulties of this time. Like most fugitives, he faced the constant worry of being captured while often spending years without seeing his family and friends. When his son died in a car crash, Biggs was unable to talk to anyone about dealing with this devastating loss. Finally, in 2001, he turned himself in and served eight years in prison.

Joanne Chesimard, also known as Assata Shakur, was a member of the Black Liberation Army who was wanted for bank robbery. In 1973, when stopped by the New Jersey State Police, she killed one of the troopers, was caught, and was given a life sentence. Other members of her group took prison guards as hostages and helped her escape from prison. Five years later she fled to Cuba and was granted political asylum. This counts as a successful disappearance but it was the political aspect that made success possible. Several years later, the US government appealed directly to Cuban leader Fidel Castro to return Shakur to the United States, but several prominent politicians, including Representative Maxine Waters, appealed to Castro to reject this appeal.

According to the *Guinness Book of World Records*, an American, Leonard Fristoe, previously held the distinction of having the longest escape on record. He was convicted of killing two law enforcement officers in 1920. He escaped from prison in 1923 and for forty-six years lived under the name of Claude R. Willis. His freedom ended in 1969, when his son turned him in to the authorities. Because there are explicit dates ending the flights of Ronnie Biggs, Frank Freshwaters, and Leonard Fristoe, it is possible to discern a record for their years of living as fugitives. There are many others for which no such records exist. They may have disappeared, but it is possible that death was the decisive factor ending their flights.

That cannot be said for John Patrick Hannan, who at the age of twenty-two escaped from Verne Prison in Dorset, England, in 1955. Although British authorities state that he was a fugitive for more than sixty years, there is no certainty regarding the details of his escape or the validity of assumptions that he returned to his native Ireland. Dorset police appealed to Hannan to contact them and eventually ended their search. It would be interesting to know how a young car thief could have managed to hide for such a long period but nothing was heard of him after his escape.

The same doubts apply to John Anglin, Clarence Anglin, and Frank Morris who successfully escaped from the notorious Alcatraz prison in June 1962. The official account maintains that the three escapees probably succumbed to the dangers of San Francisco Bay which had long been regarded as the most important feature of Alcatraz. While optimistic friends and family members express belief that the three men fled to South America where they lived happily ever after, there are no reliable reports of their having been seen. They have been the subjects of many television programs about missing fugitives.

In a similar fashion, Jerry Bergevin escaped from a Michigan prison in 1969. In 2013, when Bergevin would have been eighty years old, the Michigan Department of Corrections called off the search. Another successful escapee was Glen Stark Chambers who killed his girlfriend and was given a life sentence in 1975. He convinced other prisoners to place him inside a crate and load the crate onto a truck. He hid in the truck as it left the prison and was able to jump off before the driver noticed him. The most recent effort to find Chambers was in 2009 when the sister of his victim made a public appeal. Glen Stewart Godwin was another murderer who managed to escape from California's Folsom State Prison and flee to Mexico. In Mexico, he was convicted of drug trafficking and sentenced to prison in Guadalajara but was able to once again escape. In spite of a $100,000 reward for information leading to his arrest, Godwin has never been seen again.

Another important case that demonstrates the difficulty of hiding from authorities who are determined to catch you is that of Victor Figueroa who walked out of a minimum-security prison in upstate New York. As soon as his absence was noted, the police launched a full-scale search. He was never seen again, not because his flight was successful but because he apparently fell into a mine shaft near the prison. He certainly escaped from prison but his disappearance was merely an unhappy alternative to prison. This and most other cases are clear indications of the severe challenges faced by criminals who endeavor to disappear.

The importance of having someone in law enforcement making a case his personal crusade is often demonstrated with the conclusion of a fugitive's long flight from justice. There is an unlimited number of felons to be apprehended but a limited number of law enforcement officers. In 2021, British authorities reported on the capture of Ambrose Nicholas O'Neill, one of Britain's most notorious fugitives. O'Neill was known as a vicious thug who was skilled at breaking out of prison. His time on the run began in 2007 and over the next thirteen years he lived under assumed names. His police nemesis was constable James

Gill of the Nottinghamshire Police, an officer who was commended for his personal dedication to the cause of bringing O'Neill to justice. PC Gill devoted many hours of his off-duty time examining reports about O'Neill's activities, and developing leads that helped bring the case to a successful conclusion. While PC Gill is a recent example of how a single officer's dedication can defeat a fugitive's efforts to disappear, he is one of many such officers who have devoted their efforts to apprehend a notorious fugitive.

Fugitives often credit their success not to skills but rather to the system's indifference about people whose crimes are relatively unimportant. The heavy caseloads of law enforcement personnel work in their favor. As noted elsewhere, minor criminals are more likely to be apprehended by accident or their simply being in the wrong place at the wrong time.

Funding Your Flight

Unless you are blessed with a personal fortune or connections to something like the "rat line" employed by Nazis at the end of World War II, you will face the burden of raising your own money. Many fugitives of modest means have been forced to utilize criminal methods of fund raising. Some, such as the Symbionese Liberation Army (SLA), robbed banks in order to stay ahead of the police. The photograph which exposed "captive" Patty Hearst as an active member of the group was taken by a bank surveillance camera during a 1974 robbery of a San Francisco branch of Hibernia Bank. As she brandished an assault weapon on the bank employees it became apparent that she was no longer an unwilling captive. The public was no longer sympathetic to her. As a result of this, the SLA and Patty Hearst became the most wanted fugitives in the US. When captured, she was sent to prison where she remained until being pardoned by President Jimmy Carter.

One fugitive who was still on the run described his need to rob a bank to fund his escape in the early days. He went on the internet to get tips on how to do that and on key points to consider in selecting a bank. From the internet, he learned the best time of day—when there were the most customers in the bank—to rob it. It was also best to rob a bank that was in a strip mall so you did not have to race across a large parking lot to get away.

Less flamboyant methods of financing are preferable to an action that would bring your picture into every post office in the country. This approach is far removed from the tactics of a common thug or a desperate person who uses a gun to take money. It requires a desperate fugitive who is also cunning and sophisticated. Many fugitives are highly educated individuals who have found themselves in a difficult situation, forced to fund a new identity but not wanting to become a "common criminal." One of the more attractive methods used by cash-strapped fugitives is identity theft. It does not require violence; it brings a reasonable financial reward; and it does not place you on the list of America's most wanted. It does not force you into that category of felons who have used a firearm in connection with a criminal act. Synthetic identity theft is the term used to describe recent tactics in stealing an identity. With synthetic identity theft, a thief will take one component from one individual and others from two other people. The advantage of this tactic is that it is more difficult for authorities to track the felon. This method can also be used to take a huge assortment of identities by stealing bulk data from a medical office, for example. The market value of that much data is considerable and selling it does not endanger the perpetrator.

An indication of the subtlety of identity theft is the success that some hackers have had by using children's smart toys to steal personal information. Because many toys connect to the internet and collect considerable information about children, they can become an avenue for a hacker to gain access to important information. A few years ago, a Fisher Price toy bear which had a camera in its nose could be used to intrude into your family's WiFi. The V-Tech toy was an especially notorious example of an unsecure toy that was used by hackers. The joy of young children blessed to have such a toy was short-lived, as parents learned from the media about the criminal misuse of the toys.

Dichotomy of Fugitives

I HAVE ALWAYS BEEN AMUSED WHEN I would hear people speculate that Osama bin Laden was hiding somewhere in a cave in Afghanistan. I never believed that to be the case, and none of my former CIA colleagues did either. Can you imagine how uncomfortable it would be to spend one night in a cave? It's cold. It's wet. There are no amenities. It's like camping, but darker, colder, wetter, and less fun. Then imagine doing it for years. Nobody could accomplish that. And I don't care how "battle hardened" they might be. With that said, most fugitives seek out comfort. Why live in a cave when you can just as easily hide in the middle of a major city? Why hide in a cave when you can live in a shack in the mountains? Even better, why not live in an apartment four blocks from the beach in Santa Monica, California?

For many people, living in the wild, surrounded only by animals and insects may be very hard and even impossible to achieve and maintain. That is why it may be much better to hide in plain sight as was seen in the sixteen-year flight of James "Whitey" Bulger. Hiding in plain sight is a very dangerous alternative as facial recognition technology and artificial intelligence in general, are becoming more and more of a reality in our lives. This technology can be used by law enforcement authorities, as well as by ingenious hackers who have been enlisted to find you. If you are simply an innocent person who wishes to have a brand new life, leaving the former one behind, this aspect may not be a problem, but otherwise the need to find some place where you can't be located is especially urgent.

An important characteristic of these fugitives is that they were the object of searches which specifically targeted them. They were not random targets, but people whose notorious activities had made them prominent in the law enforcement community. There were often huge financial rewards for information leading to their capture.

As can be seen from some of the above cases, the identity of fugitives is an important consideration. Those who have access to resources, like Joseph Mengele, can enjoy a lifetime of success. For those with only modest resources or those who might be impoverished, it is more difficult.

The cases cited above represent a degree of success. In many ways, however, they are exceptional, often because they were exceptional individuals who enjoyed advantages in making good on their escape plans. The case of Eddie Maher, whose autobiography *Fast Eddie* chronicles his seventeen years on the run from British authorities, is a good example. In 1993, Eddie Maher, who had been a driver for the Securicor courier service, helped rob a Lloyds Bank of $1.6 million. Maher took his share of the robbery—an estimated $160,000—and fled to the United States. Since this was well before the 2001 terrorist attacks, security checks were less stringent and he easily made his way to Dallas, Texas, where he met up with his girlfriend, Debbie Brett, and their son, Lee. Although his girlfriend had not been aware that

Maher was a fugitive, she agreed to stay with him and to start a new life in the United States.

Eventually, the family settled in a small town in Colorado that was not known as a tourist attraction, thus reducing the likelihood of their being spotted by people who might have read about Maher's well-publicized escape from British justice. Drawing from their criminal proceeds, they paid cash for a modest four-bedroom house and began to integrate themselves in the community. Maher had a false passport in the name of Stephen King, and Debbie, who had no passport, identified herself as Sarah King. Maher secured a driver's license under the name Stephen King and used this to open a bank account. In 1994, the two traveled to Las Vegas and got married under false names, thus acquiring a marriage license which helped them in supporting new identities. He had a forged birth certificate for their son, Lee, so the child could enter school and apply for a social security card. They did not have a green card and as a result could never have a legitimate job that would provide a stable income.

As a result of this difficulty, Maher decided to move to New Hampshire and create a new identity to replace his Stephen King persona. He built the new identity based on that of his brother Michael who had married an American woman and lived in the US for several years before returning to the UK. Using his brother's personal details, Maher contacted INS and explained that he had lost his green card. INS raised no questions and mailed him a replacement card which finally gave him the ability to have a real job. After securing a commercial driver's license, he got a job as a truck driver. In 1997 the couple had another child and Maher got a better job as an installer with Nielsen TV systems. Although Debbie was forced to work in menial jobs which informally paid cash, Maher did well in his work and got a promotion that enabled them to move to Florida and later to Missouri.

Their economic circumstances failed to stabilize and, as a result of excessive credit, they were forced to declare bankruptcy in 2010. While the bankruptcy process requires disclosure of a great deal of information, it did not expose Maher and Debbie as frauds and they began to rebuild their financial lives after this. While his planning over almost two decades was successful in establishing and maintaining an assumed identity that was never exposed by the authorities, it was his own son who was responsible for their eventual apprehension. His young son, who had become increasingly rebellious and suspicious about his parents' lives, had married a girl with whom he argued a great deal. At one point, he had shared his suspicions with his wife who did an internet search that revealed their criminal background. In February 2012, she went to the police and denounced her husband's family. Maher was extradited to the UK and served two years in prison.

Although Maher enjoyed seventeen years of success in maintaining his new identities, he was eventually exposed and sent to prison. With that, we should conclude that while he was eventually brought to justice, his efforts indicated that a person with modest resources could effectively disappear. In an interview after he got out of prison, Maher observed that "you can live off the grid in plain sight if you are not silly." The key, he said, was not to try to enjoy an extravagant life style.

An even more impressive demonstration of how to elude law enforcement is seen in the case of James "Whitey" Bulger who lived as a fugitive for sixteen years. The most common observation about his long flight from the FBI was that he hid in plain sight. In spite of this relatively successful flight from justice, there was nothing romantic about the Whitey Bulger story. His first arrest took place when he was just fourteen years old. By all accounts he evolved into a vicious killer who personally killed or ordered the killing of those who got in his way as he took control of the Winter Hill Gang, which was Boston's most notorious Irish-American gang.

That triumph took place in 1979 and was made possible by the fact that Bulger had been working as an FBI informant since 1974. His relationship with the FBI was cemented through his friendship with

FBI agent John Connolly. Like Bulger, Connolly was a South Boston resident who went to school with Bulger. Bulger's FBI connections, facilitated by Connolly and one other agent, ensured that he was able to avoid arrests by the FBI. This arrangement was successful until 1994 when the Boston Police Department and two other law enforcement agencies launched an investigation of Bulger. Suspecting that Bulger had informants in the FBI, these local agencies did not inform the FBI of their investigation. While Connolly could not protect Bulger from the impending arrest, he did warn him that local authorities were on their way. This began Bulger's sixteen-year flight from the FBI. Although Bulger, along with Osama bin Laden, made the FBI's Most Wanted list, people joked that he was really on the FBI's least wanted list because they could not find him.

The key to Bulger's success was that he did not flee to an exotic locale where he could enjoy a lavish lifestyle. In fact, he and his girlfriend took up residence in a modest apartment building in Santa Monica, California. Bulger's criminal tradecraft was exceptional. He rarely left the apartment except to occasionally feed stray cats. His girlfriend told neighbors that her "husband" was suffering from senile dementia and could not interact with people. They paid for everything—including rent—with cash from his approximately $800,000 stash which was kept in a wall along with an impressive assortment of weapons. Although he was an avid book lover, he stayed out of bookstores because he suspected the FBI would be sending his picture to as many bookstores as possible. Eventually, authorities, realizing Bulger was too illusive, focused on his girlfriend and conducted surveillance on any places she might visit. This provided the first convincing clues that eventually led to his capture.

Bulger's capture took place on June 22, 2011, in the basement of his apartment complex. He had been drawn to the basement by a call from building maintenance that someone had tried to break into his storage room. Acquaintances observed that the eighty-one-year-old Bulger should have suspected that this was a ruse to draw him away from an apartment that contained weapons. It appeared that Bulger, who did not resist or even deny his identity, had lost the energy required to maintain his secret identity. Following his arrest and conviction, Bulger was sentenced to life in prison. For him that was only eight years because in 2018, while confined to a wheelchair, he was beaten to death, apparently by another criminal who felt a special animosity toward Bulger who was widely regarded as a "rat." While Bulger was the FBI's target, it was through his wife that law enforcement found him. The FBI knew that Mrs. Bulger liked to get her hair done in salons every week. So they circulated her photo to salons all around the country. Finally, an employee of a salon in Santa Monica, California, responded that Mrs. Bulger had had her hair done in that salon. The FBI put a surveillance team on the salon, spotted Mrs. Bulger, followed her back to her apartment, and caught her husband without incident.

There are certain fugitives who fit into a special category and are able to make their flight a comfortable experience. In 1972, a much easier time for people on the run, Robert Vesco demonstrated how this could be accomplished. Vesco was a sophisticated criminal who was involved in stock manipulation while making illegal contributions to President Richard Nixon's reelection campaign. He went to a very comfortable sanctuary, Costa Rica, and proceeded to make strategically planned donations. Most importantly he donated more than $2 million to a company associated with the president of Costa Rica and established himself as a valuable asset to Costa Rica. The Costa Rican government demonstrated its appreciation to his contributions by passing a law that barred Vesco's extradition. Today this would not be so easy to accomplish because the United States has extradition treaties with most nations. Eventually, because of political controversies, Vesco was barred from Costa Rica. Medical problems became the final step in his flight. In 1982, Vesco relocated to Cuba because it could provide treatment for his severe urinary tract infections while protecting him from extradition to the US. Eventually, his criminal activities aroused the suspicion of the Cuban government which charged him with drug smuggling in 1989. In 1996, he was indicted for fraud and illicit economic activity prejudicial to the state. In 2007, it was reported that Vesco had died of lung cancer in a Havana hospital but many people suspected that he faked his death.

Political fugitives, as opposed to politicians, can be more effective at hiding. During the 1970s the Weather Underground was responsible for a series of bombing. By 1976, they were responsible for twenty-five bombings, including targets such as the Capitol Building, the US Department of State, and the Pentagon. The group's small size and its use of guerrilla tactics made it difficult for law enforcement to foil its bombing campaign. While some members were captured, others used false identities and spent decades on the run. Another terrorist, Sara Jane Olson, who was a member of the notorious Symbionese Liberation Army, spent twenty-six years living as a housewife in Minnesota. Political refugees are more difficult to find because their fellow refugees are unlikely to give information about them to law enforcement. This is different from common criminals whose motivations are not idealistic but are more often based in greed. The common criminal is much more likely to turn in a fellow criminal.

According to retired FBI agent Joe Judge, white-collar fugitives are usually less effective when forced to take flight in an effort to avoid prosecution. Some, like commodities trader Marc Rich who fled to Switzerland after being indicted by Rudy Giuliani, are eventually forced to rely on political connections rather than their own skills. Because his wife had been very strategic in making high level political contributions, including to the Clinton Foundation, Marc Rich was able to secure a presidential pardon from Bill Clinton as he was leaving office. His eventual survival was not a result of effective tradecraft, but effective political connections.

Jacob (Kobi) Alexander and Sholam Weiss demonstrated the importance of developing political connections and understanding legal technicalities in making good on their flights. Alexander was an Israeli-American business prodigy who was charged with wire and securities fraud. He went to Namibia because it had no extradition treaty with the US and proceeded to ingratiate himself with Namibian elites by providing scholarships for promising young people who wanted an education. At the same time, he provided housing for low income Namibians by constructing solar-powered buildings. With these undertakings, he created a reputation for himself as a generous, reform minded guest of the nation and was able to enjoy a public, comfortable lifestyle. Eventually, Alexander tired of fighting against extradition and agreed to a guilty plea under which he received a thirty month sentence. He was transferred from a New York prison to an Israeli prison by virtue of his Israeli citizenship.

Sholam Weiss ran a successful plumbing supply business in Brooklyn, but was charged with financial fraud. Weiss fled to Brazil where he immediately acquired a regular girlfriend while also enjoying the attentions of numerous prostitutes. Apparently, he knew that under Brazilian law, a person could not be extradited if he had fathered a child in Brazil. Apparently losing confidence in his ability to impregnate a girlfriend, Weiss went to Austria where he was arrested. In his subsequent trial, Weiss fared poorly and received the longest white-collar sentence on record—835 years. Like Marc Rich, his salvation came in the form of a presidential pardon by President Donald Trump on his last day in office. This remarkably long sentence is generally explained by Weiss's refusal to accept the plea deal offered by prosecutors. The National Association of Criminal Defense Lawyers initiated an appeal to President Trump on Weiss's behalf.

Impact of Special Circumstances

A VARIETY OF SPECIAL CIRCUMSTANCES CAN CHANGE the dynamics of the situation for those who have taken flight, as well as for those who are pursuing fugitives. Many regions of the world are in chaos because of civil war, famine, economic collapse, or natural disasters. Civil unrest, such as the riots that took place in the United States in 2020, is another factor that will change the equation for anyone who is living off the grid. A skillful fugitive will be able to calculate the impact of such circumstances on his flight and sometimes can identify conditions that will ease his disappearance.

Regime change, something that is often turbulent and violent, can create difficulties for law enforcement or others who may be searching for a fugitive. The collapse of a regime may often be accompanied by open borders that are not supervised by authorities. The massive movement of millions of people at the end of World War II represented opportunities for those who were fleeing the Soviet Army, people who were being victimized by anti-German ethnic cleansing, or common criminals who sought to benefit from the chaos. Many others were simply seeking a new life further west and saw their chance to escape the postwar environment or unhappy personal circumstances. New identities were easily available by picking documents of the dead who littered the countryside. In almost any German city, the names of missing people were posted on the now destroyed buildings that were once their homes. Not surprisingly, thousands of fugitives made good on their efforts to escape justice or to flee from persecution. Many Eastern European refugees were able to join labor service battalions and work on American or British posts.

When the French government was destabilized by student revolts and national strikes in the late 1960s, the frontier between France and its neighbors was completely open. While the government was focused on political turmoil, simple matters of criminal activities were less compelling. As French prisons fell into disorder, many prisoners fled and were able to leave the country. Frontier checkpoints were often completely unmanned and travelers freely passed into or out of France.

The collapse of Eastern Europe's communist party states created opportunities for smugglers, human traffickers, or others who were, for whatever reason, on the run. The Hungarian-Austrian border only functioned on the highways, so thousands of people were able to leave the road and walk into Austria. From there, they went to the Federal Republic of Germany where they could begin new lives with the support of FRG services. It was in this period that thousands of young women were pulled into human trafficking thinking that they were being taken west to find opportunities.

More recently, the massive reorganization of almost everything about our lives has changed the situation for fugitives as well as those who pursue them. The emergence of COVID-19 as a public health threat in 2020 saw radical changes in how things work. Most importantly, lockdowns limited movement while there was a major increase in surveillance justified by efforts to limit the spread of COVID. Given the difficulty of finding hiding places, some fugitives simply turned themselves in. Health care has always been a concern for fugitives, especially as their success enables them to reach old age and suffer from the normal ailments of the elderly. Several years ago, Clarence David Moore, a longtime fugitive who, at age sixty-six, arrived at the Franklin County, Kentucky, sheriff's office in a wheelchair telling authorities he wanted to put his criminal past behind him. With the emergence of the COVID pandemic health requirements are more urgent and being a fugitive in need of testing or vaccination means a direct encounter with health services that are cooperating with law enforcement.

Even more compelling is the fact that a fugitive's movements are dictated by COVID restrictions. While his experience or tradecraft might dictate certain actions, those actions are either more difficult or simply not possible. With curfews, the most convenient time to move—nighttime—might ensure a random encounter with police. During the UK's spring 2020 lockdown, police apprehended a record number of three hundred fugitives, most of them tripped up by COVID-related security measures. One fugitive was noticed simply because he was not wearing a mask. Also, as people were forced to rely more on technology, the ability of law enforcement to monitor internet communications became a major factor in capturing fugitives. Air traffic was subject to closer scrutiny than ever before and this enabled British authorities to capture the sixteen-year fugitive Mark Fitzgibbon as he flew in from Portugal to attend his birthday celebration in his hometown. According to Julia Clegg, a private investigator in Canada, more fugitives are turning themselves in, especially if they are in areas most affected by COVID and less covered by effective medical services.

Many authorities claim that increased information sharing and use of border management monitoring technologies have increased the ability of law enforcement to apprehend fugitives who might otherwise have remained free. The development of programs such as contact tracing—the effort to determine who has been in contact with whom—have also been effective in pursuing fugitives. CCTV-based surveillance and facial recognition technology are less useful because most people are now wearing facemasks. As a result, there is now greater reliance on programs that build a profile that will help analysts understand personal vulnerabilities. By knowing that many fugitives will be alarmed if they believe loved ones may be facing a health problem, authorities can predict when those people may attempt to make contacts with parents or children. There is now a greater emphasis on machine learning, artificial intelligence, and geo-location tools.

The successful fugitive is forced to move with some regularity, lest his hideout be compromised. In the COVID environment, movements are limited, thus forcing the fugitive to stay longer in a familiar environment or with his longtime associates who normally provide shelter. Because of the limits imposed by COVID restrictions, the risk of capture is now greater because it is more difficult to find a safe haven.

Another COVID related problem will emerge as a result of the increasing requirements that every person take the vaccine. That requires more than simply taking the vaccine. You must be able to prove that you had the vaccine. What is under consideration is creation of what is, in effect, a "COVID passport" that will show the date and place of your vaccination. It is likely that you must have this if you hope to travel to another country and may well need it even for domestic travel. This is a requirement that will impose another layer of documentation and will create another vulnerability that can trip up a fugitive. The increased authority of public health officials means that there is yet another service that might identify a fugitive.

Other factors, such as police forces distracted by their public health responsibilities, staff shortages, new and more demanding safety protocols, and the almost universal wearing of facemasks, have created opportunities for the more innovative and imaginative fugitives. Smart fugitives know to avoid the use of airplanes in the pandemic and look at slower options such as cargo ships. If they must communicate, they are more likely to employ encrypted communications services such as Telegram or Signal. Smart fugitives also understand that if they are using a digital form of communication, every keystroke is recordable, so they have no illusion of privacy. The economic devastation of the lockdowns has meant that many people have become desperate to return to their homelands or, as a minimum, go in search of other economic opportunities. These movements have greatly increased the stress placed on certain border controls. As a result, border police have less time to consider criminal activities such as smuggling or catching fugitives traveling on fake passports.

Practical Steps in Disappearing

All too often, disappearances are a mystery because of the difficulty in determining the motives of the person who has gone missing. While a person may have been seen on CCTV, it is impossible to determine motives for the individual's movements. The person may be using their cell phone, but in many cases technology will not reveal the nature or content of the conversation. Even in a small town, an individual may be followed through CCTV with only a few interruptions. Yet, eventually that person may disappear from surveillance, never to be seen again. What follows is often speculation about whether the disappearance was deliberate or if it is the result of kidnapping or misadventure. Of course, an investigation about what the individual might have taken when he departed can provide some clues without yielding absolute certainty. It is clearly a disappearance but you cannot be sure if it was intentional as it would be for those who are actually fugitives.

The first steps you take as you set out to disappear can have a great impact on the ease with which you are followed. As long as you are not an escapee from prison, it will take the police several hours or even days to begin their pursuit. Your first step might be to contact a friend whom you have not seen recently and ask to stay overnight with him. You should obviously offer a casual reason for this request rather than saying the police are after you.

As an alternative you might travel more than fifty miles from your home and stay in an upscale motel. Police are less likely to raid a nice place whereas the typical "flophouse" will be treated with less courtesy. Do not pay with cash because that might attract too much attention. Instead of cash, use a prepaid credit card which is equally protective of your identity. You should leave your car in a busy parking lot, perhaps one at an airport. Hopefully, the police will waste time checking to see if you have actually taken a flight somewhere. Have an airport taxi take you downtown, where you can locate a limousine service to take you to your next destination. Do not hire a limousine service at the airport since that will provide too much information about your plans. What you should be looking for at this time is a high end destination that is not regarded by police as a likely hiding spot for a person on the run.

The most obvious way in which most people can be easily tracked today is through their cell phone. Thus, the first step in any effort to disappear is to avoid the use of a smart phone. Through a multitude of aps, people allow tracking devices to record their location and cell phone providers are usually willing to allow authorities to locate you. By abandoning your phone on a train or a bus, you can mislead those who might be using it to follow you.

Even more damning for the individual who may plan on disappearing is the network of social media through which people provide details about the most private aspects of their lives. Even the most naïve of those hoping to disappear know to terminate their accounts on Facebook, Twitter, Instagram, or any of the other platforms. However, it is equally important to avoid using social media in order to keep track of old friends. Even with a new profile and name, a careful observer can detect who may be watching known associates of a person attempting to flee his old life. Your efforts can also be undermined if someone has been carefully following your online posts for a long period of time. If they have done so, much of your personal information is already known. The best you can do is look at that information yourself and consider how you can change your behavior so it does not reflect past preferences and interests you may have posted on social media.

Another effective tactic for the person who hopes to disappear is to conduct a search for what might be new residences. You may find an attractive community and do an apartment search which requires a credit check. The establishment of a local bank account will add to the fiction that this is to be your new home. Meanwhile, after creating this paper trail, you continue your search and visit multiple sites to further confuse any searchers. The prospective fugitive should be careful in his selections of a new home. If he is an urban resident, it might be tempting to look for a small town but such a location carries its own dangers. People in small towns notice newcomers and are automatically curious about them. This makes exposure of their true identity much more likely. One of the best places in which you might settle is a town that has a large and consistent transient population. College towns or towns experiencing sudden growth are preferable. Otherwise, a mid-sized city is a better location since a newcomer will not stand out as much there.

One seemingly innocuous factor in selecting an apartment for a long-term stay is the presence of a washing machine. You will need to wash your clothing with some regularity. If you have no washing machine in your apartment, you are forced to visit a laundromat. Laundromats are places that bounty hunters frequently check in their efforts to identify fugitives. Staying in your apartment is important if you want to avoid being seen.

In hiding your true identity, it is important that you change yourself. This means dropping your old hobbies and habits. You should reconsider how you dress, what sort of foods you eat, and what you do for entertainment. If you have long been an avid bowler, take up golf. By placing a pebble in a shoe, you can even alter how you walk, so from a distance you look different.

In conducting the search for a new home, it is important not use a personal computer. Going to a library or an internet café is the easiest way to avoid detection by someone who checks your search history. Using a neutral computer and immediately erasing your search history, you can avoid the most common risks of a computer search for a new home. A site such as Justdelete.me is a useful tool in eliminating traces of your cyber existence. Be especially careful not to check you email while you're on the run. There is no surer way to tell the authorities your location than by checking your email. It might be tempting. But don't even think of trying it.

It is equally important to avoid encounters with the police. Avoidance of routine traffic stops in which you may have been guilty of some offense is particularly dangerous, especially if there is a warrant against you. Even parking tickets or broken taillights, innocuous by themselves, can pose a major problem. An encounter with the police is equally important because it could constitute an unhappy occasion in which you might learn that the new identity you have purchased actually has its own criminal record. That would mean that you could find yourself in jail for a crime which was not your own. It is also likely that during a period of incarceration the authorities could learn your true identity so you can then be charged for a crime you might have committed.

Any encounter with the police will likely require you to present various forms of identity. While an individual can secure the necessary documents, there are industries that provide this service, although they may be costly. For the economy-minded refugee, a search of death records or even a tour of a convenient cemetery offers a good start. By getting the death certificate for someone born reasonably close to your date of birth, you can secure items such as a passport or a driver's license. Opening a bank account can be accomplished with these items in your possession.

Most complicated disappearances come with an expensive price tag. Therefore, when preparing for this, a person needs to give careful thought to how they can either get money or be sure they can gain access to the money they already have. Since credit or debit cards may be flagged, depending on the urgency with which the refugee is sought, it is vital to have either cards under another name or cash that can be used for routine purchases. Anonymous credit cards which are not tied to a name are one vehicle that might be used. The danger with this type of card is that, since it is not linked to any name, it can be used by anyone who has it in his possession. It can be used without providing any identification.

There is a common assumption that faking one's own death is one of the best ways to disappear, although experts generally advise against this. This is an action that brings a variety of its own problems. More important, unless you are trying to escape legal action, it is not illegal to go missing. Each year thousands of adults do this and face no legal consequences. In fact, police are reluctant to follow up on reports of an otherwise healthy adult who simply leaves. A more severe problem with this is the likelihood of attracting the attention of insurance fraud specialists. While leaving is legal, the manner in which a person leaves might open someone to allegations of insurance fraud or some other activity supporting an elaborate faking of one's death. Of course, there are simple ways of faking your death if you can travel to a place like the Philippines where you can purchase a fake death certificate with a statement that your body was cremated. So much of the world is characterized by blatant fraud that this is an attractive work-around for an otherwise intractable problem.

Another challenge to an effective disappearance is the proliferation of closed circuit television cameras. It is generally estimated that there is one camera for every ten Americans, so the likelihood of your being picked up by facial recognition software is very high. A person in an urban area might find his image captured at least sixty times each day. Attempts to alter your appearance might be of some help, but the growing sophistication of this technology means that this threat will always be a factor in your escape plans. There is no easy way past this threat except to avoid malls, franchise restaurants, and other modern and high-end shops.

Of course, one of the most difficult aspects of disappearing is that you must leave behind the things that you treasure, most importantly family and friends, and maybe even a monetary fortune. Efforts to contact those people will quickly expose you to those who might be searching for you. Even things like a favorite car or special souvenirs you have collected may undermine your efforts at disappearing.

Practical Steps in the Worst-Case Situation

THINKING ABOUT DISGUISES, MONEY, OR TECHNOLOGY that might help is a useful process. But it is equally important to plan for the absolute worst case. What do you do when all else fails? Exotic locales are beyond your reach and you are completely on your own after living in a world of codependent humans. You have two immediate concerns. One is to escape from the forces that are tracking you. The second is to get away so you will not be a burden to those about whom you care because the pursuit teams will be looking at them. So how to you plan for the worst-case scenario? And, in fact, what is the worst case?

In short, when all else fails, a fugitive may have to head into the wild. There are numerous cases in which fugitives have done this. Rather than slowly and carefully fading away, these people literally have to jump out of windows in the middle of the night as their doors are broken down. Most of these stories do not end well. It is vital to consider how you might survive if hiding in plain sight is not an option. Not all wilderness areas are under the same legal jurisdiction. There are some lands where your presence, even if seen as purely innocent, would arouse the suspicion of people like the park rangers. Any person who includes flight into the wilderness as a possible option should know if their innocent presence would be legal in that particular area. It would be embarrassing, to say the least, if you had fled prosecution for murder or bank robbery but were tripped up by a trespassing charge.

The four basic necessities for survival under these harsh conditions are water, shelter, food, and warmth. The first step is to acknowledge that even though you can survive, you are going to suffer. This psychological preparation can help you in your first steps. For the fugitive, this means that you will not be able to share your location with anyone else and there is no reason to even think about carrying anything like a satellite phone for emergency communications. If you have been forced to head into the wild, the emergency has arrived and there is no easy return.

If you prepare for such an eventuality, you might designate a primary contact with whom you can communicate during this time. A general assumption is that the more people who know about what you're doing, the greater the chance of exposure. But, keeping that in mind, you may benefit from having a contact in whom you have total trust and confidence. You need to be sure this person is trustworthy, as well as discreet. Necessary communications should be conducted through burner phones that both of you will have. You should use foreign SIM cards with the burner phones as they will create difficulties for surveillance teams. If for any reason you need to rely on email, use an encrypted account through ProtonMail, Hushmail, Mailfence, or Lavabit that cannot be connected to you and was created exclusively for your escape. Your primary contact must be a person who shares your outlook and is sophisticated enough to avoid compromising your escape plans. Once you have deployed into an appropriately remote location, you will not speak with your primary contact and all the burner phones will be abandoned, perhaps on public transportation so their eventual discovery will not leave a clue about your location. As you prepare for your flight, it is helpful if you consider which of your forms of communication are trackable. Using those, you can casually offer hints to your associates that you if you go away, your plan is to go to a certain place but be sure to mention places that you know you will never use. This will create a false trail that will mislead a pursuit team and give you an early advantage.

Preparation of what is known as a "bug out bag" or "go bag" is an important step in your preparation. This is not the essential logistical components for your journey into oblivion but the items you will immediately need simply for your escape. One of the essential items is cash because you have to avoid electronic purchases that might expose you. In addition to cash, you need things that you can use for barter should you find yourself in a place where money is not the most important value. A lock pick kit is something that could be placed in the "bug out bag." With sufficient practice, the lock pick kit can help you get out of difficult situations. A handcuff key is another useful item. If you have packed tools such as knives these can be exchanged for assistance or services from people you meet.

The Psychology and Skills for Survival

IT IS IMPORTANT THAT YOU UNDERSTAND your situation in terms of where you are, who is pursuing you, and your prospects for survival. If you think you are being pursued, you should find a location that offers concealment but allows you to observe the movements of the pursuit team. While you know there is a need for special equipment, the first tools that you can use are your senses of hearing, sight, and smell. Every environment has a certain feel or pattern—sounds made by birds, for example—and you need to be alert to any changes. It there are crickets and they suddenly stop making noise, it is likely evidence that someone is moving into your area. If there are other people in this area, you should understand their normal movements and behavioral patterns.

You should also be alert about your physical situation and check to see if you have been injured in any way during your headlong flight into the wilderness. If you have injuries that have caused bleeding, they need to be treated immediately. Even a minor loss of blood will impair your ability to function. An important factor in maintaining your health is to keep shots or vaccinations up to date for any regions in which you might find yourself. Bandages and disinfectant can be applied when you recognize the need but shots must be taken in advance of a health crisis. At the same time, you need to be aware of the equipment upon which you are depending for your survival. If some of it has been lost or damaged, this can undermine your chances of eluding your pursuers.

Most people in a stressful situation believe they need to do something, anything! They are reticent about simply waiting while doing nothing. This is a dangerous tendency if you are a fugitive running from a pursuit team. Acting in haste will reduce planning time and result in loss of equipment or self-awareness. Failure to notice factors such as a change in temperature can lead to difficulties as your environment changes. Becoming disoriented can be a fatal mistake that could lead to your capture or death. One of your more important tools is your ability to read a map and to pick out identifiable terrain features so you don't get lost. It is important that you have a map and a compass. These are essential tools for your land navigation efforts. You can, of course, gain a general sense of your location and direction if you know how to study the moon and the stars. Your starting assumption is that the sun rises in the east and sets in the west. If you have the benefit of shadows, you can use them to determine both the time of day as well as directions. The shadow tip method will give you a more precise indication. With this method, you take a stick that is about one yard long, clear a spot on the ground, and position the stick so you have a clear shadow. The first shadow mark will be west. Wait fifteen minutes and mark this second spot on the ground. By standing with the west mark to your left and with the second mark to your right, you will be facing to the north. By using the moon and the stars, you can find your direction during the night. You can also study the moss on a tree to determine directions. While moss may grow all around the tree, you simply identify the part where its growth is the thickest.

Almost as important as knowing your location is being aware of where your pursuers may be. By following their movements, you can determine the direction in which they are moving and learn more about their likely plans. You should be able to determine their rate of movement to calculate how you can best stay out in front of them. If you cannot see them, it is possible that you will hear them.

An especially important psychological factor for a fugitive is dealing with isolation. For the fugitive, isolation is more than being alone. It means being away from the comforting things of your normal environment. A person who suddenly finds himself outside his comfort zone and lost in an alien location is going to fill the impact of his loss. This can be debilitating and will have a negative psychological impact that will be physically draining and depressing. It can eventually even be fatal if the fugitive loses hope. The best hope is if the fugitive realizes that he will face this situation and will be better able to avoid panic. By developing coping methods for dealing with this, the fugitive is more likely to remain calm in this difficult environment.

The case of Shoichi Yokoi, a Japanese soldier who remained in hiding on the island of Guam after World War II illustrates the necessary attributes for survival. While hiding in the jungle for twenty-eight years, he never lost faith in his cause and believed that his comrades would come for him. When he was discovered by local hunters in January 1972, he still felt that he was in danger and was determined to fight to avoid capture. Although weakened by years of malnutrition, he resisted the hunters who, he feared, would make him a prisoner. In his culture, to be taken as a prisoner was seen as a disgrace.

In his account of his experiences, *Private Yokoi's War and Life on Guam, 1944-1972*, Shoichi Yokoi tells how he and his fellow soldiers survived this long experience. In order to elude the enemy, they carefully erased their footprints as they moved through the jungle. In the early years of their flight, they lived by killing local cattle for food. When this source of food was eliminated and the Japanese soldiers feared the encroaching American troops, they were forced to move even deeper in the jungle. Their food source became poisonous toads, river eels, and rats. Driven by necessity as well as by a need to remain occupied, Yokoi learned how to use reeds to build a trap for catching eels. He constructed a shelter which he fortified with bamboo cane. In the last eight years, he was alone because everyone else had died. Even when he suffered from severe illness, he swore that he would not allow his corpse to be found by his enemies so he crawled back to his underground shelter. Shortly after he encountered the hunters, Yokoi was given a hero's welcome back in Japan.

George Ray Tweed, an American Navy radioman in the Navy Communications Office, had a similar wartime experience. Tweed was also in Guam during World War II and, after the surrender of the US garrison in 1941, he evaded capture by the Japanese for two years and seven months. Tweed and five other men escaped into the jungle rather than surrender and become prisoners of war or to be executed, as other Americans were. The Japanese offered rewards for information leading to the capture of any American but, because of his radio skills, the reward for Tweed was ten times as large as the normal reward. As a skilled radioman, Tweed knew how to make the most primitive radio work. In March 1942, he managed to repair a Silvertone radio and use it to receive broadcasts from San Francisco. When the Silvertone radio died, he found a Zenith Electronics radio. With this, he was able to hear the USAFFE, "The Voice of Freedom," broadcasts from US forces on Corregidor Island. Unlike Shoichi Yokoi who had no idea that the war was over, Tweed was informed about the progress of the war. The information from these two stations helped maintain the morale of Tweed's group, as well as of the local populace which was helping them. For a short period of time, Tweed even published an underground newspaper known as the *Guam Eagle* by using an old typewriter and carbon paper.

When his group was finally apprehended, all of them were executed except for Tweed who eluded capture. One reason he remained strong was that the locals with whom he worked told him that as long as he was alive and free, they believed that the Americans would come and rescue them from the Japanese. The Japanese tortured and executed locals whom they suspected of aiding the Americans. During his remaining time on Guam, he kept occupied by studying algebra and making shoes for the family that provided shelter for him. In July 1944, Tweed used his skills with a mirror and semaphore to contact two American destroyers that were preparing for an invasion of Guam. The destroyers sent a boat to rescue him and Tweed, who had used his time to collect information about the Japanese positions, was able to provide valuable intelligence for the invasion force.

Taken together, the skills demonstrated by Shoichi Yokoi and George Ray Tweed will shape your psychological outlook. They can bolster your confidence and your belief in yourself. If you become rattled and lose confidence, you will make mistakes that could result in your capture. Because most of us have become accustomed to an easy, comfortable existence, a loss of self-confidence will make it increasingly difficult to use the skills, tools, and advantages that you have. Retaining a belief in yourself shapes a positive mental attitude and helps you retain the will, something that is essential to your survival. Apathy and helplessness in the face of your problems will have a negative impact on your outlook. The same is true for any feelings of bitterness about your plight. The realization that you have lost your old life can be overwhelming, especially if you had enjoyed that existence. On the other hand, if your previous situation was less than desirable, you may feel that your difficult new life is preferable to the alternative.

Preparing for Survival

ALTHOUGH YOU REALIZE YOU HAVE TO pack essential supplies, you must understand that having too many supplies will be a problem. An early step in this process is determining how you will carry your survival kit. After all, if you are going to have to carry all the gear and it weighs you down, your survival prospects will also go down. This means that you have to set priorities for what you will carry. If you find yourself in a bitterly cold environment, your first concern must be shelter. Unless you are hopeful about finding a cave, which I would certainly not recommend, you will need to carry something that can be used as a tent. A cave might be convenient, but use of the cave requires you to remain in one spot. If you have a tent that you can carry, you will be able to keep moving and stay ahead of your pursuers. This is your only long-term survival plan. Packing a compass, something which is light, could be helpful if the terrain is difficult and you face the prospect of getting lost. Given the danger of navigating in the wild, your emergency pack should include first aid supplies. At a minimum, you need gauze and other items that would enable you to stop the bleeding from an injury. Consistent blood loss will not only weaken you, it could lead to your death within just a few minutes. An ointment like Neosporin is easy to carry and can provide pain relief and disinfectant.

Medical concerns will become paramount should you be injured or become ill. This is one of the most difficult issues you will face and, unless you have some medical training, it represents a threat to your survival. The least you can do is to be aware of the locations of drug stores or a medical clinic. While a hospital might be nice, hospital employees are always alert for patients who might be running from the police. Most hospitals will have a police officer or security personnel in the facility. They will notice any suspicious injuries and will certainly ask you about them.

Another useful thing, if your planning supports it, is to preposition supplies that you assume you will need but cannot physically carry. If you expect to flee into a certain area in the wilderness, you can place a cache in relatively secure places. Therefore, if you realize you are running out of an essential supply like food or water, you know where you can find some without traveling into a town where you might be seen. You can also use the cache for basic medical supplies that you need. Three or four such locations can be the difference between success and failure.

A multipurpose knife is an essential item. It will enable you to cut wood for a fire as well as to capture food. If you catch a squirrel, you don't want to have to tear it apart with your hands as you prepare it to eat. Having a lightweight sleeping bag is also helpful and can protect you from freezing to death if the temperature falls. The most likely cause of death in wild is hypothermia rather than starvation. Lacking a sleeping bag, it is important to remember that dirt is a good insulator.

By burying yourself in dirt—or pine needles, fallen leaves, or any other forest debris—you can preserve your body warmth and survive until the next day. A reflective thermal blanket can provide protection that you will need if you must contend with severe heat. To survive freezing temperatures, you need to be able to start a fire, so matches would be useful. Of course, wet matches are worthless, so you need an alternative. A flint serves that purpose, can be used repeatedly, and is lightweight. During your time in the wilderness, if you should fall in a stream or get wet because of rain, the prospects for death by hypothermia increase greatly. Your only chance for survival will be to start a serious fire within five minutes to rescue yourself from hypothermia.

Prolonged exposure to cold will result in exhaustion, depression, and loss of the will to live. As you build your shelter, be sure it is large enough for you to lie down in comfortably. Be sure it is not too large because a small shelter is more effective in retaining your body heat. It should be sturdy enough to keep out wild animals who might want to join you and to survive should things like small branches fall from the trees. The shelter needs to be in a position less likely to be spotted by pursuit teams and, if such a team discovers you, it is vital that you prepare an escape route. An irregular shape will make it harder to spot. It also should not be erected close to a stream that is likely to rise above its banks and flood your shelter.

There are many types of shelters and your selection should be governed by the materials you have and the amount of time you can put into construction of the shelter. One of the easiest to build is the poncho lean-to which requires a poncho, a few feet of rope, and two trees that are about six feet apart. By attaching the ropes at knee level, you can lower the profile of the lean-to, making it less likely to be seen by searchers. Another shelter, the poncho tent, provides a lower profile and offers protection from the elements on both sides. With this, you will be warmer, but the structure of the poncho tent is such that your escape time is slowed down if the pursuit team finds you while you are resting. Should you find yourself in cold, snowy terrain when you do not have a tent, a snow cave is an attractive alternative and is often warmer than a normal tent.

When you study all the equipment you plan on taking, it is vital to consider the function to be performed by each tool. As you look at each tool, think about what that tool does. If one tool can perform both functions, limit your pack to that single tool. Redundancy is often a good principal in casual planning, but it is a bad idea if you are operating in a dangerous wilderness environment where you have to carry everything yourself. You will need a case to carry all these items and the case should be waterproof, durable, and attachable to your body either by a strap or with clips for your belt.

The symptoms of dehydration are low energy, headaches, dizziness, muscle cramps, and, eventually, unconsciousness. If you are to survive for any length of time while relying on your own resources, you need to have a consistent, natural supply of food and water. Rain water—whether falling from plants or from the sky—can be trusted and standing water can be made potable through the use of purification tablets. A metal container, if available, can be used to boil water and kill most of the bacteria. If you are relying on this, be sure to being the water to a rolling boil and maintain it for as long as possible. Water is a more urgent requirement than food and three days is the maximum time for survival without water. In addition to rain, you can collect dew by pressing clothing onto the ground. By dragging a piece of cloth behind yourself, you can also gather water that might have been collected on plants. If you notice ants going up a tree, you can search the tree for the water that might be collected in a groove in the tree.

A banana tree is a reliable source of water. You can get to this water by cutting down the tree and leaving a stump that is about one foot high. The stump will provide water for four days. Any plant that has a soft, moist center can also be used to get water by cutting off a section of the plant and squeezing it until water comes out. The roots of most plants are another source of water. If you pull the roots out of the ground and cut them into small pieces, water will come out when you smash the roots.

Plants that produce nuts and berries, some mushrooms, and edible roots can be a source of food, if you are able to distinguish between ones that are safe and those that are not. You can be spared the prospect of starvation if you have access to these plants. This is a time when your survival may well be dependent on your ability to distinguish between poisonous and edible plants. Many people believe you can distinguish between poisonous and non-poisonous plants by watching which ones the animals eat. While that observation has some merit, there are certain plants that are poisonous to humans but not to animals. Another assumption is that plants that have a red color are poisonous. The truth is that some of them are but others are not. The point to remember is that there is no single and simple rule that works for all plants. You have to study a botany manual to be sure.

Mushrooms are difficult to identify with precision and mistaken identification can be disastrous. They are especially treacherous and some can result in sudden death from liver failure. Often there is no known antidote for a poisonous mushroom. In countries where it is common for people to hunt wild mushrooms, fatalities are common. Mushroom poisoning can affect either the gastrointestinal or the central nervous system. It is important to avoid unnecessary touching of unknown plants because contact dermatitis will be persistent and will spread by scratching. If the plant's oil comes into contact with your eyes, this will be especially uncomfortable and will hinder your ability to move in the forest. If you are exposed, try to remove the oil with soap and cold water. If water is not available, wipe your skin with dirt or sand.

The results of being exposed to a poisonous plant range from minor irritation to death. One means of exposure is eating the plant. If you cannot distinguish between types of plants, this can easily happen. You can also be exposed by simply coming into contact with certain plants. Most people are familiar with poison ivy and know that it is easy to stumble into this in the woods. People can also be exposed to poison ivy if they make the mistake of burning it in their search for fire wood. Inhaling the fumes of a poisonous plant can result is severe injuries, especially in the throat and lungs. People have varying degrees of sensitivity to poisonous plants. If you have previously been exposed to poison ivy, you may be immune and will not suffer from exposure. On the other hand, some people are extremely sensitive and have debilitating reactions when exposed.

In the right season, you can probably find edible flowers but remember that some plants become toxic after they wilt. The edible inner bark of trees is an abundant source of energy survival food and is relatively easy to find. It has a bad taste but it is easy to harvest and will keep you going. By using your knife or even a sharpened stick, you might manage to catch fish and small animals that will be a reliable source of nutrition. Small animals, of course, are fast and difficult to catch. If the animals defeat your hunting efforts, the most reliable alternatives would be insects. They are easier to catch and are nutritious. In many cultures, they are consumed in much the same way as we might eat potato chips.

In general, animals are less likely to become a food source and more likely to represent a threat to your survival. Common sense will alert the fugitive to the threat posed by large and especially dangerous animals such as bears or any creature that has horns. The small animals that may look like attractive targets in your hunting endeavors can actually be very dangerous to encounter. Their fangs and claws can have a deadly impact. Small venomous snakes are especially dangerous and their small size lulls unskilled woodsmen into complacency about them. Even bees can be deadly as many people suffer from allergic reactions caused by their sting.

We recognize insects because they have six legs and arachnids because they have eight legs. These small creatures bite and sting. As a result they are serious pests. Barring an allergy to their particular toxin, they are rarely fatal. Even the most dangerous spiders, such as the brown recluse, rarely result in sudden death but can cause a painful, long-term illness and tissue degeneration around the wound. By exercising caution and checking your footwear and bedding, you can avoid these dangers. Most people recognize ticks and know they spread diseases such as Rocky Mountain spotted fever and Lyme disease. If a tick attaches itself to you, it must be removed within six hours. Otherwise, it will be able to spread the disease organism into your body. By using a common insect repellent, you can protect yourself against ticks.

When calculating the dangers you may face, remember that each environment has its own special threat. Rivers, for example, can be a danger when crossing them, not simply because you could drown, but by the pests that use rivers for their homes. In North American rivers you need to be cautious around turtles because they will bite when threatened and, as a result, you can easily lose fingers and toes. In rivers in South America, there are electric eels and piranhas which have sharp deadly teeth and can be found in shallow water. If you find yourself in saltwater, you will face an assortment of dangerous fish. One is the rabbitfish which Pacific natives who are skilled in their handling of it view as food. Lacking their skills could result in a fatal injury. Off the Gulf Coast of the United States, you may find the toadfish which has sharp toxic spines along its back and can be painful for anyone who encounters it. The poisonous scorpion fish can be found in the Indian and Pacific Oceans and sometimes in the Atlantic. Its sting will cause intense pain. There are also other fish that can be deadly if eaten. This includes the blowfish, the barracuda, and the triggerfish.

If you are determined to carry a weapon, it is important to remember that mere possession of a gun can trigger suspicion. Much of this will be a function of the environment in which you find yourself. If your goal is the have the largest possible number of rounds, the .22LR is probably the best weapon to carry. Otherwise, the most important concern is to know what you are protecting yourself against. The biggest weapon is not always the best weapon and the most versatile weapon is preferable. The .22LR is probably easier to find and much cheaper than the larger weapons. There are some regions in which carrying a weapon is routine and will not arouse suspicion. Other areas are just the opposite. Sometimes one of the best weapons to have is something that can also be viewed as an ordinary tool. A screwdriver does not automatically generate suspicion but it can be an effective weapon. It is not conspicuous and can be easily hidden beneath basic clothing. It also fits into the category of an everyday carry item.

Survival Movements in the Wilderness

If you have been driven into the wilderness out of desperation, your most important consideration is survival. A fugitive who has planned in anticipation of this worst-case scenario should have basic information about the region where he has taken refuge. That planning should have identified actions to be taken in emergency situations because when faced with a crisis, there is no time to formulate various courses of action. As is so often the case when attempting to disappear, planning is the key to avoiding capture. This is the time to have planned options to disrupt the movements of the pursuit team. Your knowledge of the terrain will help you identify likely routes the team might take. Knowing those trails will give you an advantage if you can block the pursuer's passage or slow them down. While you cannot afford to engage the pursuit team as you would an enemy combat unit, you must recognize that this team if your adversary. If you are going to be in a situation requiring you to cross a national border, you will need to study possible locations which are easier to cross. If you are planning while you have computer access, Google Maps is a good resource. You should identify remote roads having limited traffic and few apparent border patrol checkpoints. When you find yourself on the ground in that area, basic reconnaissance is essential if you hope to find vulnerabilities such as there being only occasional police presence. If you are on foot, you have a good chance of crossing into another country without being detected.

If you have found possible areas in which you can hide, you should go to them and stop all movement, if possible. Unless the terrain is especially treacherous, movement at night is usually preferable because it is more difficult for the search team to see you without night-vision goggles or heat-sensitive equipment. As you move, avoid roads and trails, stay away from man-made structures, and do not approach settlements. If you have selected a sufficiently remote area into which you intend to disappear, there should not be many people permanently living there. During this time, slow movements are best and it is important to occasionally stop to listen for sounds of pursuit. Your route to this location should not follow a straight line but should offer occasional radical changes in direction to more easily throw off pursuers. Once you arrive in the spot you have identified as a hiding place, camouflage yourself in the most secluded area possible. You should look for the thickest vegetation in the area and be careful not to cut it lest that provide a clue for the pursuit team. This is the time and place for absolute silence.

While in the hide site, do not build a fire or prepare food because the smell of food will travel for a great distance. It is important to vacate the site after twenty-four hours and move to a different site. Carefully clean the old site so you do not leave clues for the pursuers. Your next move should be into a more secure and isolated location where you can "hole up" for a longer period of time. This will be an area in which you can prepare food and rest. It should be near a good source for water.

The Use of Disguises While Disappearing

As a matter of routine, we take care in deciding how we wish to appear to other people. Questions such as manner of dress or fashion, hair styles, our method of walking, or even our speaking style are basic. Even a person who gives no thought to actually disappearing must consider these issues. It is important that we determine the type of impression we make or if we hope to make no impression at all.

In creating a disguise, people generally hope to do one of two things. First, they may want to make themselves less recognizable or, second, they may attempt to look like someone else. If that person is a specific individual—such as a famous person—the task is more difficult than simply trying to appear to be a businessman or an athlete. A simple disguise is preferable to an elaborate, complicated disguise. If there are too many elements to the disguise, it is more likely that one component may fail, thus exposing you.

Some of the most basic elements of a disguise are weight gain or loss, adding or removing facial hair, wearing a wig, and wearing glasses. It is also helpful to use contacts that will change the color of your eyes. Shoes lifts or dental inserts can do a lot to change one's appearance. Adding age lines or artificial scars can distract an observer by drawing his attention to a prominent feature. He would notice the scar while paying less attention to the target's overall appearance. There are basic components that the fugitive can carry in order to rapidly change his appearance. A roll-up hat, facial hair that can be glued on, a mirror, scissors, and paper towels can help transform your appearance. A voice recorder is useful if you need to create a fake accent and want to practice it. Most specialists suggest that plastic surgery, which is expensive, offers limited advantages and is not worth the trouble. Indeed, there are many examples where famous fugitives—John Dillinger may be the most famous—had plastic surgery and it did nothing to throw off their pursuers.

The manner of physical deportment is an important feature in recognizing a person. One of the most famous World War II spies was Virginia Hall. Because of an accident, her leg was amputated from the knee down and she walked with a limp. Working in occupied France, she hid her limp by disguising herself as an elderly woman who walked with a shuffle. It is also possible to change your gait by feigning a limp so at a glance you do not look the same. Some specialists, however, argue that you should never attempt to change the way in which you walk unless you are in a dire situation. The limp is difficult to maintain over the long term.

If possible, you should have a disguise kit that contains cosmetics, hair dye, and small items that will help you transform your appearance. Makeup specialists claim that you can create a light disguise in ten minutes while it will take thirty minutes to make a major change in how you look. The most important aspects of a face in terms of the accuracy of identification are hair, hairline, and upper part of the face. There has long been a debate about which is more important: precise and distinctive facial features or the entire face. The most consistent assumption is that adults are more likely to focus on the entire face as a whole. Some people, especially children, employ a *featural encoding strategy* and will more likely cite specific descriptors when trying to remember a person. Older people are more likely to rely on a holistic strategy which means an overall description. In spite of this, it is true that changing just one facial feature can transform the overall appearance.

Disguises will vary according to the length of time they are to be worn. If your disappearance begins with a single incident in which you do not wish to be identified, you have a different set of possible options. If it involves a criminal action such as a robbery intended to provide funding for your disappearance, your disguise requirements can be satisfied with a temporary covering. The goal is to conceal all or even part of your face during the criminal action. During this, the culprits may conceal their faces with a ski mask, floppy hats, stockings that are worn on the head, or simple masks that might be used on Halloween. Presidential masks have been among the most common coverings. Research has demonstrated that if a culprit is wearing something as simple as a hat, it changes the facial composites seen by witnesses. All of these have the important advantage of being something that can be removed in seconds which is crucial given the fact that you cannot hope to walk down the street wearing these disguises. For that part of your escape, you need to look as normal as possible since your objective is to blend in with other people.

These short-term disguises are totally different from ones that need to change your appearance over a long period of time. For those circumstances you will need to rely upon a relatively simple arrangement that embodies, as noted above, changes in weight, altering facial hair, wearing some sort of glasses, or fashion changes. With effective makeup kits, a person can dramatically change their pigmentation and perhaps pass themselves off as a person of a different ethnicity. This is a disguise that must be continuously worn during specific periods of your time as a fugitive. It is how people are likely to see you on a regular basis even if your life is that of a recluse. At some point, you will have to come outside. If your appearance is inconsistent, you are likely to become an object of suspicion.

In considering the creation of a disguise, there are three things that have the greatest impact and can be altered with relative ease. These changes relate to hairstyle, facial hair, and adding or removing glasses, even sunglasses. Tinted eyeglasses are significantly more effective than regular glasses since they more completely obscure the eye. With that accomplished, the memory of a possible witness will be thrown off. An important consideration of these relatively superficial changes is that they are useful as a short-term measure to avoid being identified. They have limited long-term utility.

When a person begins planning to bolt from his current existence, a considerable degree of planning is required. In addition to research about places to go and how to get there, it is useful to give some thought to disguises. An effective disguise will have a positive impact on the fugitive's survival prospects. Fortunately, there is an abundance of professional assistance available for the fugitive. Thanks to modern technology, much of it taken from film studios, a person's appearance can be radically altered in terms of age, sex, ethnicity or many other features. For as little as ninety dollars, a person can buy a professional disguise kit on line by going to a spy shop like https://www.daytonaspyshop.com. This kit uses human hair, just like at the CIA, to create a fake mustache and goatee. Each hair is looped into a mesh lace device that can be fixed to your face using spirit gum. Other sites such as https://www.thecostumer.com rent a variety of devices that will allow a fugitive to change his appearance. They offer a large selection of hundreds of costume wigs, false eyelashes, fake eyebrows, fake beards, and costume moustaches. Still more expensive are the silicon masks provided by https://www.realfleshmasks.com.

Other more expensive providers boast that they can create disguises that look exactly like other people, including famous people if the fugitive has a need to pass himself off as an important figure. It is more likely that a fugitive might want to change his ethnicity and this is possible by utilizing technology that supports the creation of East Asian, Indian subcontinent, indigenous American, or a variety of other ethnic identities. It is even possible to create a disguise for a person suffering from serious disfigurements although such a disguise might attract too much attention. Of greater value are the disguises that allow a person to change their apparent age from young to old or the reverse.

Considerations about an effective disguise are especially relevant when a fugitive knows he might fall under suspicion and have to deal with possible witnesses. It is important to calculate the odds of being recognized and understand strategies for avoiding identification. If a fugitive is being pursued by law enforcement, he should anticipate tactics used by the police in trying to identify him. If the target is in police custody and under suspicion, there are two procedures police can employ. They are (1) simultaneous and (2) sequential lineup procedures. In the former procedure, the lineup members all come out at the same time. If the fugitive has developed an effective disguise, his chances for surviving the simultaneous lineup are better. With the sequential lineup procedure, each lineup member comes out individually. While this procedure seems more dangerous for a fugitive, studies have indicated that there is little difference in the accuracy of identifications. The most important factor in both procedures is that when a fugitive has changed his appearance, identification becomes more difficult.

Practical Steps for Maintaining Safe Contacts

ASSUMING THE FUGITIVE IS NOT WILLING to live in a cave devoid of all the comforts of civilization, it is vital to determine relatively safe ways of accessing needed funds and maintaining some contact with individuals from the old life. To avoid social networks and internet-based contacts is the first and most obvious step. These safeguards are vital to having the low profile that will create challenges for pursuers. However, any kind of life requires some means of support.

Modern technology has created almost insurmountable problems for individuals hoping to enjoy secret lives far from the troubled lives that caused them to flee. However, there are a few modern technological features that actually help those hoping to live off the grid. Prepaid credit cards and virtual phone numbers are two of these features.

It is obvious that your old bank cards will be useless in your new life. Therefore, a first step in creating a personal support mechanism is the utilization of prepaid credit cards or store gift cards. A person who never uses any sort of plastic will attract attention and may raise suspicion. A prepaid credit card is an alternative that enables one to make reasonable purchases and appear to fit into the local community. The worst that someone might think is that you are a person with marginal credit who cannot get a regular credit card. At best, you might be taken for a frugal person who wants to avoid the interest payments associated with a regular credit card.

Even with a new identity, it will still be necessary to make phone calls. Since your cell phone or a normal land line are not safe, prepaid phones and calling cards are the best alternative. These can be purchased at almost any convenience store and will, of course, be paid for in cash. The prepaid phone can be traced but as long as the phone is not in your name, it provides a measure of security otherwise absent with your old cell phone. Safety is enhanced if the person whom you wish to contact also uses a prepaid phone. If you call your established contacts at their recognized number, law enforcement authorities will see that an unknown phone is calling and immediately be suspicious about the caller. The resulting phone surveillance could provide important clues as to your location. Even if you have removed the GPS device from the phone, you can still be tracked by identification of the cell phone towers you have used.

There is another frequently helpful innovation that improves security with a normal smart phone. This is to use a foreign SIM card when you are in your own country. Such a SIM card can be purchased online and, by using it at home, you will undermine the normal tracking procedures that work quickly when you use a correct SIM card. While the cost of this is minimal, it can help a fugitive gain a small advantage that will help him elude pursuers.

The so-called burner phone has features that make it useful for a person who has been forced into hiding. It is a no-contract, prepaid mobile phone that does not support a GPS or any smart phone apps that might expose you. Today, surveys indicate that 70 percent of Americans believe the government uses personal information for more than just tracking criminals. A person who watches the news will realize that in this uncertain political environment, the government may have an unhealthy interest in religious communities, social clubs, or political movements. After the attack on the US Capitol, authorities spread their surveillance net further than ever and former CIA director John Brennan has suggested there is a need for an anti-terror program for use in the United States against its own citizens. He compared a broad segment of the American population to ISIS and argued that it was just as much of a threat as a foreign terrorist group.

While the term "burner" was associated with the activities of criminals who wanted to avoid police surveillance, it has many uses that are legal and above reproach. If your business involves making lots of phone calls, you will want to keep your personal phone number out of your commercial activities where you are likely to be called at all hours. A person who uses an online dating service might want to keep his number secure in order to protect himself from unhappy dates that are angry when he did not call them again.

The most important aspect of the burner is that it protects your privacy. If you anticipate that you may need a burner phone, it is important that you make that purchase well in advance of its likely time of use. If you are being investigated, law enforcement can more easily check for recent purchases if you elect to use only one. Even better, buy several burner phones and just throw them away after use.

It is important to note that there are limits to the utility of the burner. If you call a known associate, phone surveillance will show that a burner phone was used to contact one of your known associates. If that person is under surveillance, your burner phone can reveal patterns of your behavior without showing your true identity. When the burner phone contacts those who are known as your regular contacts, hunters will identify a pattern and will know it is you. If, however, your associate uses a burner phone it is more difficult to catch you. If the burner phone uses an Android system, it can be easily tracked because it has a built in WiFi and forces you onto the internet. If you do not connect to the internet, you cannot be easily located. If you do connect, you will be easily located. It is also true that a burner phone—just like a smart phone—can be tracked by cell phone towers. By using three cell towers, police can triangulate to determine the location of the phone. A few years ago, a Syrian journalist was reporting on government atrocities. She used a burner phone to contact her publisher, thus allowing the government to locate her position. Shortly afterward, an RPG blast hit her office and killed her. If you are being tracked by the NSA, *voice printing* technology can be used to recognize your voice no matter how you communicate. The most important factor in this area is the identity of the agency that is trying to locate you. If it is not the NSA, voice printing is probably not a factor that will undermine you.

Another weakness of the burner phone relates to how you use it. If you use it on a regular basis and leave it on even when not in use, a pursuer can examine your call logs and trace the number back to you. With that, you are exposed and likely captured. If you are in a situation that requires use of the internet and you do not have the benefit of a VPN, you phone can be traced. How you buy the phone is also crucial. As noted elsewhere, you should always use cash and make the purchase from a vendor who does not know you. Because some countries require cell phone vendors to collect personal information from all their buyers, this could be a problem. If faced with that problem, you will need to have someone front for you in making the purchase. If you are able to do this, you can avoid part of the trap but will have created a witness who might identify you. It is equally important how you dispose of your burner phone when finished with it. The key rule is that the phone can never be traced back to you. If you place the phone where someone will pick it up and travel with it, authorities will have a more difficult time catching you.

Many burner phones, such as Alcatel A206, are cheap and easy to use. In its standby mode, the Alcatel A206 battery will last up to seventeen days. The same battery will support five hours of actual talk time. The burner phone helps the fugitive because it enables a person to have a number that will be used for selected important calls without revealing your location. While you may be in area code 540, the burner phone will show a different area code. It will also enable you to make a call to someone who may have blocked your number. Upon completing those calls, the user "burns" the number by discarding the phone. An especially clever person might leave the phone where it will be picked up and used by a stranger whose journey will create a false trail. A reality television show called *Hunted* recently demonstrated how authorities could utilize modern technology to locate a fugitive. The ten participants on this show were promised a generous prize if they could elude pursuers for twenty-eight days. Contestants could shelter with friends, hide in plain sight in a big city, or escape to the mountains and live in a tent. While the hunters were able to use dogs, drones, or helicopters, their most important weapon was their ability to construct profiles of the fugitives that helped predict their behavior.

While most people likely use the burner phone for purely legitimate reasons, many governments are considering a ban on burner phones. While authorities may complain about criminal activities, such a ban is more likely to be used as a way of suppressing unsupervised communications. Violent protest movements also utilize burner phones so they can communicate with each other while organizing against the police. Proposed legislation would require formal registration of the burner phones rather than their absolute elimination.

A practical variation on the burner phone is the burner phone application that allows users to transform a smart phone into a burner phone that will employ a false number. If you need to call someone but do not want them to see your actual phone number, this device will enable you to do that. Calls made to your burner phone can be transferred to your smart phone without breaking the anonymity of your smart phone number.

A burner phone app has several advantages over the regular burner phone. The regular phone will usually provide you with only one or two temporary numbers. By contrast, the burner phone app will enable you to draw from a much larger pool of numbers. You can buy as many as you need and dispose of each number after you use it. This way, you do not create an anonymous phone number identity that can be studied by those who are pursuing you and, in the end, undermine your anonymity. With the burner phone app, your real number and your temporary number are on the same device. You do not have to carry an extra phone—or two or three extra phones—and deal with the problem of disposing that phone in a way that does not transform it into evidence that might be used against you. When you finish with the temporary number, you simply delete it. The burner phone app gives you all the temporary numbers you need without burdening you will an additional device. You can still use your smart phone as you would normally if you are in an appropriate situation.

While there are many burner phone apps from which you may select, not all offer an infinite supply of temporary numbers. One of the best known apps is Phoner but there are others such as Hushed which offers numbers from forty countries but requires a constant Wi-Fi connection. Another is CoverMe which boasts of its "military grade" encryption and offers a "private vault" which can be used to store documents or any sensitive information. One shortcoming is that CoverMe can only be used if five countries. The Burner system is another highly rated app but can only be used in the United States and Canada and will supply you with only one number at a time. If your plan requires regular use of more than one number, the Burner system will not work for you. TextMe Up is yet another app that can be used. It has the benefit of being simple to use and is the only system that will provide you with one number that is absolutely free. With

TextMe Up, you buy a subscription for whatever time you might need, varying from one week to one year.

As noted above, Phoner is the best-known app and is praised for its flexibility and reliability. It offers phone numbers from more than three dozen countries and regularly adds others. It operates on a pay-as-you-go credit system which enables you to purchase just as many credits as you require when you assess your situation. It also features a consistent customer service team that is ready to deal with any problems that might arise. With Phoner, even the numbers on the text messages cannot be traced. There are, however, some complaints as customers have stated that they lost phone credits when they made no calls and that the customer service was not helpful. They were directed to Google Play which insisted they could not refund the customer so the next step was to contact the developer. But most users have expressed satisfaction and describe the app as being even better than promised.

An important innovation that will support the effort to maintain communications which are relatively secure is the virtual phone number. This is also known as direct inward dialing and is a telephone number that is not associated with an actual telephone line. The virtual phone number is programmed to forward incoming calls to a customer selected preset telephone number. There is an important element of deception with this system in that it shields the actual location of the caller. There are some suspicious elements of the arrangement since it enables a firm actually located in Russia or in China to look as though it is really located in an American city. During the dislocations associated with COVID which forced millions of Americans to work from home, this system enables them to have access to a centrally located office land line. Utilization of this arrangement means that a fugitive who does not want to reveal his location even to those whom he trusts can be reached through what appears to be a local number.

Although there are numerous companies offering this service, one of the best known virtual phone services is jConnect (www.jconnect.com) which works with eFax. With this service, you can use any email address in order to receive voicemail and also get faxes in your email as a PDF. This system does not require a physical location that might increase your vulnerability if you are being pursued. Equally important is the fact that more than 11 million users employ this perfectly legitimate system for their business needs.

Just as modern technology is so often the enemy of a person who has taken flight, there are significant ways in which it can help that individual stay hidden. There are numerous expensive devices that can help him undertake communications which are absolutely necessary. One of these is the DigiScan Delta 100 X 4/12G system which helps the fugitive detect wiretaps, micro cameras or other surveillance devices being used against him. For the detection of wiretaps, hidden cameras, or mobile phones, there is the Delta X 2000/6 countersurveillance system. Another valuable tool is the encrypted messenger RAW for secure communications. This can be used for phones with Android and iOS systems. Like most of the tools for creating disguises, these technological supports can be purchased online. It is important to be discreet in making such purchases because you don't want your pursuers to realize how well prepared you may be.

Searching for Secure Communications

ESTABLISHING A METHOD FOR SECURE COMMUNICATIONS is an absolute necessity for almost any endeavor. Business or personal matters depend on having a way of sharing information with the confidence that there are no unwanted listeners. Although almost no method of communications is absolutely secure, there are procedures and devices that improve the prospects for security.

The first instance of secure communication between a transmitter and a receiver took place in 1898 with Nikolas Tesla's demonstration of a radio-controlled boat. As is so often the case, wartime conditions hastened the development of technology. In the early days of World War II, Winston Churchill and Franklin D. Roosevelt were forced to communicate using a voice scrambler which was determined to be insecure because the Germans had a listening station that could intercept the calls. In response, engineers created a secret system known as the Green Hornet or SIGSALY. Unauthorized listeners would hear white noise while Churchill and Roosevelt could have a clear conversation. This system led to the development of several concepts of digital communications including pulse-code modulation.

SIGSALY was a massive device that weighed more than fifty tons and generated tremendous heat. In Churchill's war room, it was placed in a special air conditioned room labeled as a broom closet. The Washington SIGSALY was set up in the Pentagon and the US Embassy in London placed their SIGSALY underneath Selfridges department store because it was near the Embassy. The first time this system was used was on July 15, 1943. The SIGSALY was only used for the most senior-level conversations, rather than routine matters.

For any person who is trying to avoid contact with pursuers or to hide the nature of their communications, there are three basic concerns. The first is to hide the content of communications by utilizing a code. The effect of a code is to transform information into symbols. Generally referred to as encryption, this is a method which prevents an unauthorized party from being able to read data. The second concern is to obscure the identity of the people in a conversation. This can be accomplished by using anonymous communication devices such as burner phones or by using third-party systems. The final one is to prevent watchers from knowing that communication has taken place. Reporting on terrorist events or military movements often notes that an "elevated amount of chatter" indicates that something is about to happen. If pursuers become aware that their target is engaging in numerous conversations, they will anticipate some sudden move and be more alert.

Another method of secure communications is steganography. With this method, the data you wish to share with your partner is hidden inside of other data that appears to be of no relevance. You might be able to hide a telephone number inside an MP3 music file where surveillance personnel are unlikely to notice it. Terrorist-related steganography is often hidden inside a pornographic file on the assumption that an Islamic militant is less likely to view pornography.

For a fugitive, it is important to avoid using an identity-based system. An identity-based system, such as most telephones, makes anonymity difficult to achieve. For this reason, it is a system guaranteed to expose the identity of a fugitive. There is a common assumption that an anonymous communication device is less likely to be noticed simply because there are so many devices being used at any time. Unfortunately for the fugitive, with the development of systems such as Carnivore and Echelon, which can monitor communications over entire networks, this is simply not true. It is also important to remember that the object of your communications may also be subject to surveillance. For years, spy movies have portrayed situations in which members of an espionage cell communicate by their utilization of payphones. With technology from the era of World War II, that might have worked, but it is no longer an acceptable option.

Even the most basic email system—regardless of the name in which it may be registered—is a significant vulnerability. In 2021, the Hey messaging service conducted a survey of email traffic at the request of BBC. According to their survey, two-thirds of emails sent to personal accounts contained a "spy pixel." The tracking pixels are usually a file that is as small as 1x1 pixels and is inserted into the header, footer, or body of an email. By using this device, it was not only possible to tell if and when the email was opened but it was also possible to identify the user's physical location down to the precise street. This invasion of privacy was defended as a commonplace marketing tactic so companies could assess the effectiveness of their campaigns.

Maintaining Contacts in the Digital World

If you choose to live in a community where some interaction is inevitable, it might be necessary to use a phone or even social media applications. Depending on the situation that you are in, safety measures must be developed and followed.

In general, the normal user doesn't realize how much information he is giving to others through social platforms and poor security enforcement rules. Apps such as Facebook, Twitter, and Instagram serve as open windows on your life. It is equally important to avoid simple and predictable passwords and failing to effectively configure the firewall. With a good antivirus program on all your computer devices, you can prevent your life from becoming an open book. Presenting online one's daily routines make a person predictable, easy to follow and to control; posting personal pictures on the internet makes it easy to be located (be it by manual work or by an AI app); opening up infected email, choosing easy passwords based on public information—the name of a husband or a child, the name of a favorite book, etc.), and many others make it so much easier for the person or agency that is following you.

To be more precise, if the motive of running away doesn't involve criminal acts related to cybersecurity or the motive for searching the web is not obscure or easily traceable, safety measures are not necessarily that strict, but otherwise, they are highly required. In the following pages, several methods will be explained that may decrease the risk of being exposed. The intensity of your safety measures is directly dictated by the situation that you are in.

Tor is a well-known web browser that allows its users to surf the internet in a more private way, giving them access to the hidden part of the internet (websites that are otherwise blocked by the majority of browsers). In order to use it, it is necessary to download it from the official site, which is torproject.org. Essentially, this tool transmits the data from the client to a series of nodes/relays (represented by the computers of other people who have offered to help the community). It is based on a routing algorithm from proximity to proximity until it reaches its final destination—the targeted server.

Like many other methods, this doesn't offer complete anonymity to its users. Sometimes, it may even attract the attention of the law enforcement agencies as it may seem to be a potential threat or even an awareness of hackers and criminals—who may use different methods to get access to the victim's computer and use the infected device to their will—and also law enforcement agencies. Certain websites even deny access to Tor users. The main advantage would be that you would be harder, although not impossible, to catch, depending on the situation.

VPN (Virtual Private Network) servers are a mechanism that encrypts your data and sends it to a series of proxy servers from different geographic locations until it reaches the targeted server. In other words, this tool is used in order to change a real IP address to a false one, thereby creating the illusion of being in a completely different location. Compared to the Tor method, this has various advantages such as the rapid rate of transmission and data encoding. It would be advisable to use several VPNs at the same time. Unfortunately, this tactic is not unbeatable and it would also imply a monetary cost. A key part in this plan is to choose an appropriate VPN provider that would keep your data as safe as possible. (Nord VPN might be a good choice.)

A good option would also be to combine the browser Tor with the VPN technology, making it even harder to be tracked. You can start with Tor and then access the VPN. This arrangement prevents the VPN provider from getting access to your real IP. It also keeps your real identification information hidden when reaching the target but you may unable to get access to certain websites. Those are websites that are available only through Tor without the VPN coming second. This is harder to configure and may raise suspicions from the VPN company that you want to use. VPN is easier to set up and hides your IP from Tor. However, upon arrival at the targeted server, your traffic would be not encrypted. Both these methods share a common disadvantage in that they slow down the connection.

Choosing one of the methods from above would also imply the use of a search engine. Some of the best choices would be DuckDuckGo, Ahmia, Onion, or Tor search and some other lesser-known engines. It is crucial that you use a search engine that does not compromise your security.

If you need to conceal important pieces of data, it is mandatory that your data be encrypted. One of several ways to do this is to encrypt all the files on or from your computer. It is best not to use any cloud technology, but rather to store them on a digital device. If it is necessary to use some kind of online storage system, it is best not to use the most popular ones, as they might present certain vulnerabilities. Sometimes it is best to save your data in different places and under different formats if your situation allows this.

Encrypting private data from a computer or a drive is also simply a must in terms of privacy. This is one of the main steps in ensuring that your content can't be easily disclosed in any possible way. This might be a challenging task but it can be done by simply using the appropriate application depending on the storage—encrypting the hard disk, encrypting data from a cloud or a memory stick, etc. It is very hard to determine which is the appropriate tool for one's requirements but some of the best choices might be: PGP, OpenStack platform, self-encrypting drives, auto-destruction of the drive, cryptography or steganography, etc. One of the differences between using a popular product or platform and a lesser-known one is based upon the level of ease with which one could gather data about the selected target. It should be expected that these tasks can be easily performed by an officer of the law or even by a civilian who wants to investigate you.

When it comes to the operating system, it is probably best to avoid the most popular encryption programs. There are many other lesser-known platforms that can do more to ensure the security of your data. If security has a higher priority, Tails or Qubes are very attractive.

Equally important in today's digital environment is communicating through web messaging apps. As noted elsewhere, if a person is in hiding, he must not contact the people that he left behind, especially if that contact utilizes the same number as you used to call him before your flight. This may be convenient but convenience is not the friend of a fugitive. If it is mandatory to talk with people from your past, it is vital to utilize a different system. That person's phone might be intercepted, so more precautions must be employed.

There are several methods that offer some advantages. One is steganography, the art of hiding data in other files. There are platforms that cloak certain kinds of files by transforming them into different archived formats (JPEG to PNG and many others) or by simply blocking them. Instead, there other platforms—such as Flickr or even

Facebook—that can support them. Steganography can also be combined with encryption in some cases.

Another technique might be to use an online messaging app such as Signal or Telegram, which have grown in popularity as people have become suspicious of other services. It is important to understand that just shifting from a local officially-made app to an official app from another country is not necessarily a solution. Your data may be harder to access by a local police force, but somebody in another country might be able to track you and copy your data. Signal allows users to verify the identity of people with whom you communicate by creating a set of codes known as safety numbers. This is a unique Signal feature which is not available with other apps.

While Signal and Telegram have become two of the most sought-after apps in the world, authorities are raising questions about them. Professions such as law, journalism, or politics, we are told, have a legitimate need for privacy. The problem, according to many in government or journalism, is that these apps are being used by groups that were barred from using Facebook and Twitter after the Capitol Hill riots in January 2021. As a result, Signal and Telegram are being described as "hot spots of disinformation" and are likely to be targeted by government agencies suspicious of private communications of any sort. Therefore, these relatively simple methods of communication could be shut down, thus depriving fugitives of this option of sending or receiving messages. An alternative is Viber, which uses the same encryption programs as Signal and Telegram, but is based in Europe.

SecureDrop is an open-source platform that is often used by whistleblowers around the world. It is used by numerous media publications in order to receive important data from their sources. Given the number of benefits that this platform offers to users, it might serve the purpose of communicating safely with other people when you are on the run. Of course, its utility will be a function of the type of information that is being communicated.

One other problem that may arise is associated with the use of a smartphone. With the smartphone you have to share a significant amount of data. Whether it is Android, MAC OS, or other similar operating systems, they require an email address and certain permissions to access your data, such as photos, contacts, and your location. Given all the functionalities that a device like this may have, the amount of data that an application can stock about a person is enormous, and the more you use it, the more it adds to the storage. The best way to preserve your privacy is to add as little information as possible and even add details that are untrue about yourself in order to mislead the hacker or investigator who is trying to locate you. It is very hard to track a person that uses an unknown phone, even though that phone continuously sends data and signals to providers. It is important that you don't give your real identity when buying a phone subscription. It is even better if you don't have to give up any piece of data. Another important rule would be to never contact the people from your past. Their phones might be bugged and that would mean they would indirectly have access to your location. Also, if it is necessary to make a call like this, it might be a good idea to call from a burner phone, but the conversation would have to be limited to just a few seconds.

From many points of view, social media platforms have become part of the daily lives of most people. They are a source of inspiration, a way of learning about the world and relaxing at the same time. Social media offers a continuous flow of infinite knowledge and a way to keep in touch with others. Social media can easily be seen as a portal to an artificial world but this world is under total surveillance. Most of us have had the experience of being located by a person who might have dropped of our life decades ago. Even without contacting you, a person can learn about your routine, hobbies, family, and home. The person who may be cyberstalking you doesn't need to take pictures, but simply to follow certain aspects of your life, like who you are following on Instagram or Facebook, what posts you follow, who has tagged you, who is in your photos, and what places you have visited. You can be found by someone or by an AI app in a very short time. That is why it is better not to use them. If it is necessary, though, create a new fake account, access it from a library computer, and pay attention to what you post. Never post your pictures online even from a fake account. A good facial recognition app would find you eventually if pursuers are determined to find you.

It is important to remember that any data that was on a social platform is nearly impossible to erase completely. *JustDelete.me* is a tool that you can use to help delete your accounts. But this doesn't ensure the complete wipedown of your account. The platform might keep certain data about you and disclose it to the police if presented with a warrant.

In some countries, there are laws that permit you to erase any piece of data from online storage on the condition that the data that you want to erase is not of public and legal importance. However, the use of this legislation may well attract unfavorable attention from law enforcement.

As mentioned above, facial recognition systems based on artificial intelligence are becoming more and more difficult to fool. It can be a good solution to move to a country or a region where this system is not yet well installed. Eventually, these systems might become impossible to beat, but until then there are some tactics for evading them.

These tactics are generally not used by people who are wanted by the police, but who are being sought by a jealous spouse or by other personal enemies. As a person not wanted by law enforcement, you might just simply create another virtual identity (along with another email address, fake name, and a new phone number). There is a difference between ensuring your privacy and trying to escape criminal prosecution.

As noted elsewhere, when it is necessary to keep large sums of money, it would be a better choice to keep them in a private account in a tax haven country or to buy cryptocurrency such as Bitcoin. Both methods are well used by many fugitives, but the second choice would seem the safest. It would be nearly impossible to find a cryptocurrency account as opposed to a normal bank account being found out. The biggest risk with this method is that virtual currency may be unstable and your losses could be substantial. It would also be beneficial to keep some cash on your person to meet your immediate financial needs.

If you are attempting to live as a missing person or a dead person, it is sometimes essential to create a new online identity but never go back to your former identity. It is best to try to keep such an identity as simple as possible. The effort to be excessively creative generates too many points that can be independently checked. This effectively creates another weakness. Remember, "Keep it simple, stupid!"

Another useful tactic that would apply to any type of situation would be to keep a sticker or cover over your web camera, even if you are using a public computer in a library. That camera can at any time become a significant vulnerability. Law enforcement organizations and private investigators can turn the camera on at any time and use it to spy on you.

Choosing a good password is also an important element of ensuring the safety of your account. Never use the names of people whom you love or information about yourself that can be easily found or known. In addition, it is best to use passwords that are not logical in any way—just random letters (uppercased and lowercased), numbers and symbols. Furthermore, the password must be changed periodically. Sometimes, a two-factor authentication process is better. No two accounts should have the same identification data. Given the fact that remembering long series as passwords is a daunting task and nearly impossible to accomplish, you might need to use an app such as KeePassX. This app stores your identification data for every app that you use; the data is encrypted by the use of a number of algorithms.

As we have learned more about China's social credit system, we have come to recognize how effective its mass surveillance system is. Because of this, it is almost impossible for a normal citizen to hide there. The United States also has very powerful surveillance tools that limit the ability to hide from law enforcement. These tools and methods that governments use now in China, the United States, the UK, and elsewhere are expected to expand all over the world. The only thing that might reduce the power of the surveillance state is that as long as an apparatus is created and developed by a human, another human might be able to suppress or circumvent the system.

The primary reason for discussing these tactics and procedures is to show the importance of privacy and how to ensure it on different levels. These observations are just an introduction to understanding the basic steps on how to secure your privacy and maintain your freedom if you are a person intent on disappearing. One of the central features of a successful disappearance is the ability to prevent your personal information from being known by your adversaries.

Digital Fingerprints

ANOTHER IMPORTANT STEP IN SECURING PRIVACY and maintaining freedom is the avoidance of what might be called digital fingerprints. Such fingerprints are created by the numerous digital technologies that have emerged in recent decades. As people now post pictures of themselves on Facebook, Instagram, or Twitter, they provide detailed accounts of their movements and activities. Most people rely on digital platforms to communicate with each other as well as to order everything from overnight accommodations to a pizza. This also enables them to contact old friends and make new friends should they be in a new city for a musical concert or a political rally. While most of those individuals may not see their activities as criminal, in the contentious political and social environment of today these activities may well constitute elements of a criminal prosecution.

A dramatic example of this phenomenon was seen in the days immediately following the riot that took place in the US Capitol on January 6, 2021. Some participants seemed to view this as a rowdy, violent party and were inclined to post their pictures on the internet just as they would at a normal party. In effect, they were providing evidence that will likely be used in a forthcoming trial. While technology helped people get together in the days before January 6, it also helped authorities create a narrative of participants' actions, and even thoughts, during the riot.

Participants happily posted images of themselves in various parts of the Capitol Building as well as with Capitol police officers. Because such images are time stamped, and facial recognition technology augments efforts to identify the perpetrators, it was easy for authorities to prepare their cases against rioters. In addition, since smartphones provide detailed geolocation information, defendants will face severe difficulties when they find themselves in court. Authorities boasted that by following social media and other communications, they could determine who funded travel to Washington in the days preceding the riot. Digital fingerprints will allow law enforcement to create a precise timeline as perpetrators planned their attack. With the ease of connecting to social media, it was possible for many enthusiastic civilians to ferret through various platforms in an effort to identify participants and report them to the police.

While this account of the utility of digital fingerprints applies to participants in events in Washington, it is also relevant to individuals who are involved in an effort to disappear. The technologies being applied to the search for participants in the Capitol Hill riot can be focused with an even greater impact on a person who simply wants to remain hidden. The most innocuous communication with either friends or with service providers can undermine their efforts. While it is obvious that no fugitive would participate in an event that will be widely covered by national television networks, the pervasive nature of digital technologies represents a clear threat to the person who wants to remain hidden.

Much more sophisticated than what was used after the Capitol Hill riot, cyber fingerprinting is the term used to describe a process that is similar to the fingerprinting of humans in the physical world. Cyber fingerprints are groups of information that enable a surveillance team to identify software, network protocols, or hardware devices. This data can help formulate a strategy or an infrastructure map to be used against a target. The immediate goal is to collect sensitive information about the target in order to determine what technology is being used to conceal online activities. By having data intelligence about the target, it is more difficult for the fugitive to conduct online business need to sustain his off-the-grid activities.

The Challenge of Social Security

THE KEY IN DEALING WITH SOCIAL security is how to get a new social security number. Getting a new number legally is virtually impossible. People in the witness protection program are entitled to one. Or if a person could show he was being subjected to a long-term pattern of life threatening abuse, then he might be able to get one. Otherwise, it is a daunting task.

The traditional pretechnical way of getting an identity that would enable one to secure necessary documentation was to visit a cemetery. The objective of this visit was to identify a person who had died young, perhaps no older than five, and whose age was close to yours. A good illustration of how this could work was the case of KGB agent Konon Molody who was operating as Gordon Lonsdale when arrested by MI5, the British domestic intelligence service. Molody was the son of two Soviet scientists who volunteered their son to be sent overseas as a spy. The real Gordon Lonsdale had been born in Canada and emigrated to the Soviet Union, where he died at an early age. Soviet KGB agents had selected his identity as appropriate for Molody, who used it working in both the UK as well as in North America. In this work he was remarkably successful, enjoyed an extravagant lifestyle and, using KGB funds, became very wealthy. His capture in 1961 was not the result of faulty tradecraft but because of his association with the spies Peter and Helen Kroger, who were already under surveillance.

The Gordon Lonsdale transformation would not be possible today with the difficulties of securing a social security card as part of building a new identity. Now the first step in overcoming the challenge of creating a new social security identity is to verify the birth certificate of the individual whose identity you wish to assume. With the advent of electronic verification, this is more of a challenge that it was in the past.

The creation of the Electronic Verification of Vital Events (EVVE) system, which matches other documents against US birth certificate databases, is the most important barrier against birth certificate fraud. A variety of state and federal agencies use EVVE to help verify the birth certificates needed to prove age, identity, and citizenship while also meeting requirements for employment and eligibility for benefits. A valid birth certificate is an absolute necessity for entry in the Social Security system and each state's Department of Motor Vehicles.

There are eight states that do not participate in the EVVE system and that do not have provisions for electronic verification of birth certificates. Those states are Texas, Missouri, Kentucky, North Carolina, New Jersey, Pennsylvania, New York, and New Hampshire. By using one of these states, a potential fugitive applicant has good prospects for getting a phony birth certificate accepted as valid. An adult looking for a social security number will automatically face scrutiny because most people get their social security number in the days after they're born. Therefore, the applicant will have to provide an explanation as to why he was not assigned one then. One possible justification for this omission might be having lived overseas all of one's life up to this point.

Immigration documents are another important concern in dealing with social security. Such documents are covered by electronic systems. Successful navigation of this system would enable a person to secure approval of immigration documents. If you can get Immigration and Customs Enforcement to approve your documents, there is an open door to getting a social security number.

One helpful shortcut in this process would be to get a social security number from a foreign student who has no plans to return to the United States. Such a card would work for banks and credit cards. If the foreign name was unusual and inconsistent with your appearance, you could change your name legally. The pivotal factor here is that as long as the foreign student is not returning, you can safely adopt his name. If you happen to be operating near a large university that has enrolled large numbers of foreign students, this is an encouraging prospect.

Use of a Fake Identity

The production of fake identity cards has become a big business over the years and, with our increased dependency on electronic devices, the industry has experienced phenomenal expansion. Numerous computer programs claim to be able to provide acceptable fake IDs. Those programs explain that the cards are not for criminal purposes but largely to protect the privacy of their users. Upon entering a store that offers free WiFi you are often required to log in using your Facebook account. By having a fake ID, you can create a false Facebook account and thus avoid having to share your personal information with internet-based companies. Many of those companies make huge amounts of money simply by selling the personal information of their customers. It is reasonable that you would want to avoid this and the fake ID will help you do that.

Numerous internet sites will help you get you started on this process. One of the most frequently used is *Fakenamegenerator.com*. This site will generate names, addresses, social security numbers, credit card numbers, and occupations and as well as UPS tracking numbers. It can perform this service for thirty one countries, using thirty seven languages. Many people will turn to Fakenamegenerator.com if they are concerned about the safety of accessing certain websites. Many websites will market the data collected from visitors to their site, so having fake data will protect your real information. The data that the site gives a person is randomly generated so you are not stealing another person's identity but simply protecting yourself. As you begin planning for your disappearance it is important to protect your identity, lest you leave a digital trail that pursuers can follow.

The problem here, though, is that even if your motivations are a matter of self-protection, this is a criminal offense. That goes for both completely fake documents or for authentic documents that might have been altered by adding your name or changing your age. Changing the photograph is illegal and also invites criminal prosecution. Even if you do not actually use a fake ID, possession of the fake ID is also a criminal offense. Penalties for this offense range all the way from fines to ten years in prison. A person who buys a fake ID is simply adding to the profits of a criminal gang who will use that money for other illegal activities. Of course, if you are a desperate fugitive running for his life, this transgression will not likely impact your decision to employ such documents.

You might be concerned that your purchase of a fake ID is bringing business to criminals who will use those profits for other criminal endeavors. It is also true that China is one of the major creators of fake IDs that are sold in the United States. This knowledge might actually inspire some sense of guilt. Even worse, however, is the possibility that the person who sells you the fake ID might actually be a police officer. Should you be that unlucky, your flight could be delayed for an indefinite period of time.

As you examine presentations about how easy it is to acquire a fake ID, you need to be aware of the difficulties you will face in your search for this documentation. One of the most effective instruments being used against you is artificial intelligence (AI). This is a system that can study large datasets and determine what patterns exist. While it is a formidable adversary, AI only knows the data that has been given to it by programmers. It has no intuition and cannot think out of the box. Formidable though it may be, AI alone can be overcome by a determined and desperate fugitive. The greater danger is posed by the Certified Fraud Examiners who are trained to be innovative and identify connections that might otherwise be missed by a computer program. These are the people who will help AI to interpret data in the correct way. AI spots discrepancies in patterns but the CFE provides overall direction to insure that the AI works properly. Before making a commitment to using a fake identification card, you should consult the Fraud.net website in order to see what fraud protection tools might inhibit your use of this kind of card.

Another and more insidious form of fake identity is known as *sock-puppetry* and has existed for centuries. Modern technology has made this device even more effective. One of the best examples of this was created by a PhD student at the University of Edinburgh. His creation was an imaginary thirty-five-year-old Syrian American who was running a blog in which she described her experiences in Syria during the uprising against Bashar al-Assad. His creation soon had a worldwide audience that was following the stories told by his creation. It all ended when an NPR reporter's research revealed that the whole thing was a hoax. While this was an embarrassing end to the saga, for several months it gave the graduate student a voice to express his ideas about the Syrian situation. This tactic might be useful in creating a large-scale distraction, but it is unlikely to be useful to a low-profile fugitive.

Living Off the Grid

At many levels, the notion of living off the grid is innocuous and little more than a nod to the environmentally conscious. However, there is a difference between living in complete isolation in the mountains and creating a new identity that will enable you to live out in the open but simply in a new part of the country. With the former, there is an emphasis on the desirability of living in a remote location like Alaska because of the tranquility or beauty of the area. There are designers who stress that you can live in an eco-friendly home and enjoy the benefits of secure utilities and solar power. Companies such as Midland Architecture market a tree house design that is "sensitive and self-sufficient." Promoters assure you that it will be unobtrusive in a natural pastoral setting. Prospective customers undoubtedly envision this small dwelling surrounded by peacefully grazing cows rather than special weapon assault teams advancing toward a suspected felon.

Companies that have developed these innovations appeal to people such as homeowners in states like California who are tired of living on an unpredictable power grid. They also encourage people to read booklets such as *Energy Options for Off-Grid Homes: The Role of Propane in Off-Grid Designs*, rather than survivalist works such as *The Turner Diaries*.

The idea of living off the grid is often an expression of being independent, resourceful, and reducing both expenses and your carbon footprint. The internet offers a multitude of kits that will enable you got build a cabin in the woods, provide your own utilities system, and enjoy tranquility associated with Henry David Thoreau's book, *Walden Pond*. Thoreau expressed his general outlook with the statement "I went into the woods because I wished to live deliberately, to front only the essential facts of life, and see if I could not learn what it had to teach." The book is an exploration of transcendentalism that appeals to people who might, in another time, have been seen as hippies intending to drop out of society.

This is very different from the concept of living off the grid in order to not be found by either law enforcement or curious neighbors wishing to develop a friendship. This is off the grid for fugitives or those whose desire to drop out is driven by less benign motives. Often they are survivalists who are bracing themselves for some sort of Armageddon or, at a minimum, some federal agent hoping to take your guns. As such, they are not fugitives in the sense of people who face imprisonment if the police find them. Nor are they like fugitives who are running from criminals who might want them dead.

The case of Ted Kaczynski, better known as the Unabomber, illustrates the advantages of being able to live off the grid. This notorious domestic terrorist and anarchist was also a mathematics professor at the University of California at Berkeley. While a student, he was regarded as a prodigy and had an IQ of 167, higher than that of Albert

Einstein. He saw industrialization as evil and advocated anarchism based on a back to nature philosophy. Because it was an instrument of the distortion of life, he believed that all scientific research should be ended. In 1971, having abandoned his university teaching job, he moved to a primitive cabin in the Montana woods.

For seventeen years, beginning in 1978, Kaczynski used letter bombs to kill three people and injure another twenty-three. His targets were academics, business leaders, and others with whom he associated the development and the use of modern technologies which he believed were destroying natural civilization. Because he was the subject of the longest manhunt in the history of the FBI, Kaczynski's efforts to live off the grid demanded extensive knowledge and skills. Given his intellectual brilliance, it is not surprising that he developed such an effective plan for eluding authorities. It is also significance that his capture was not a result of mistakes he made but because his own brother shared his suspicions with the FBI.

At its most basic level, living off the grid means escaping from the network of services which ties us to our local and national communities. There are some aspects of this concept that appeal to environmental activists who see it as a way of reducing their carbon footprint. There are others who see this as a way to escape the reach of an increasingly powerful state. With the appearance of numerous controls and rules related to the COVID threat, many people with libertarian tendencies were drawn to this aspiration. There are, of course, numerous criminal types like Ted Kaczynski who were motivated by a philosophy and also wanted to stay away from law enforcement.

For less notorious people who are not being sought by law enforcement, living off the grid is less demanding but, nevertheless, still requires detailed, careful planning. A non-fugitive must be aware of the development of circumstances that will create trouble for him. It could be something as mundane as business reverses that threaten his financial security. It could even be marital problems which threaten his security if they are being pursued by a vengeful spouse. Fearing this situation, the threatened individual may want to abandon his current location and identity in order to rebuild his life.

An unprepared individual is likely to panic and make mistakes that make it harder for him to blend in to new surroundings. If that person has not considered stockpiling the essentials for a new life, his flight may well end quickly when confronted by his nemesis. Therefore, careful preliminary steps will include development of a list—written or mental—that will catalog indications and warning to which a person must respond. Important and possibly threatening indications could range from negative quarterly business reports, an unfavorable performance appraisal from your boss, or suspicious behavior by your spouse or some other person in your life.

The plan must include both the time and method of departure. Utilization of public transportation carries with it certain risks. If you are stuck on a train, airplane, or bus, you lose the flexibility of being able to alter your route in response to new information. If you see police at the next station where your train will stop, you are in an extremely difficult situation. The site of a nosy official coming through the train also generates a need to suddenly change direction. You are at the mercy of a carrier which is piloted or driven by someone completely out of your control and jumping out of a window is not an option.

If you are able to depart the train or bus, you will face the burden of having to carry your bags or backpack. If you have heavy items in your baggage and are moving fast, there is the danger that you will make a lot of noise because of the items you have been forced to carry. If you are forced to move in a hurry, you may have to give up some of your luggage, meaning that you have lost things you needed. Equally bad is that you may have left behind clues that indicate what your plans might be. The type of clothing could easily tell investigators if you are heading into the mountains or to a tropical location.

Using a car gives you more flexibility. You can decide the time that you will depart and will not be dependent on the schedule for public transportation. With a car, you can usually stop when you want to, take a different road, or park your car in a place where you can sleep. There is, however, the danger of a roadblock or a police inspection point. Perhaps the authorities are conducting a routine inspection for driver's license, state inspection, or to determine if you are wearing a seat belt. Not knowing, you will be tempted to turn around and head in the other direction. That move will insure that you receive an additional amount of police attention.

Be sure, if you're using a car, to avoid toll roads, which have cameras and will photograph your license plates. And whatever you do, do not keep a toll-paying transponder in your car. It will track your every move and could inform the authorities of your whereabouts. (It goes without saying that you should not, under any circumstances, use a car with an OnStar system, which literally tracks your every move and can inform the authorities of your exact location.)

A car gives you the ability to carry supplies that will enhance your independence and your flexibility. With food and other provisions you can avoid restaurants or stores. On public transportation you cannot carry the baggage needed for supplies and equipment for living off the grid. Whether you are in a city or taking refuge in a forest will determine the clothing you carry and how much of it you need. It is vital to have clothing that help you to look like the people who are around you. City dwellers, small-town people, and rural people have distinctly different styles of dress.

If your vehicle will accommodate a bicycle or a small motorbike, you can leave the highway to take a trail or some other off-road option. This is useful and alters your profile. If the authorities know you left town in a car, they will concentrate on highways, routine traffic stops, and blockades. Now that you are using two-wheeled transportation, the search equation is different and you may have been able to get out of the search area by the time police realize you have abandoned your car.

By traveling exposed, however, you face new vulnerabilities. You are vulnerable to the elements and could suffer greatly in bad weather. It is also essential to avoid traveling in open areas where locals might notice an unfamiliar figure moving across the terrain. So you must seek wooded areas and perhaps attempt to travel at night. Moving in the darkness also brings its own hazards. Whether on foot or on a bike, you can fall into a ditch or worse. You could sustain severe physical injuries that would impair your mobility. Moreover, a traveler who is bleeding is certain to attract the attention of any observers. Visiting a clinic or an emergency room would ensure that police would take notice.

If you have avoided these pitfalls and have made it to a city, there are special circumstances a little farther afield to take into consideration. It is important to avoid crowds. If your flight was influenced by the breakdown of public order feared by doomsday preppers, there could be angry crowds that might attack any apparently vulnerable individual. If you are attempting to make a record of your adventures by taking pictures or videos, you might attract both the attention and the ire of angry crowds. Thus, it is better to rely on your memory or to simply take discreet notes to record facts you think you should remember.

When encountering angry or threatening people, as a fugitive, you don't want to attract any attention. If someone attempts to strike you, do not strike back. Cautiously move off and try to maintain a low profile. If you have a gun, never show it or use it unless you face the immediate prospect of severe injury or worse. If you have a gun, you can be sure that members of the mob have even more lethal weapons and you want to avoid a shootout.

In these difficult circumstances, you want to move away at a steady pace, as if you are not stimulated by the activities of the mob. If you take flight and run in apparent terror, you will not only attract the attention of the mob but you are also likely to get completely lost. With this, it will be difficult for you to find a relatively safe harbor. You will have shown the mob that you are not only a terrified person, you might be someone who has values he wants to protect. It is important but difficult to find the right balance between abject fear and foolish bravado. You should think about how you might respond to an aggressive dog and, at a minimum, try to appear uninteresting.

There are identifiable behavioral patterns that will support your efforts to disappear. There are basic skills that will help the fugitive avoid standing out and attracting attention. These skills make it possible for you to be a forgettable person who is not much noticed or remembered. You need to understand the base line for appearance in a new environment. Be aware of how other people look and act.

As you set out to avoid running even when you are in a hurry, there are ways to appear more natural in doing this. Casually strolling with your cell phone in your hand helps you to blend in with your surroundings. It also gives you a reason to stop by pretending that you are reading something on your phone. Under some circumstances, walking a dog enables you to pause when you feel the need to do so. Of course, as a fugitive, a dog would represent a burden most of the time. You should avoid anything like an aftershave that has a distinct, memorable smell. It has long been recognized that our sense of smell is one of our strongest senses and will most effectively prompt our memory. It is also important that you do not appear threatening. There are now ordinary looking backpacks that are bulletproof. If is important that you not wear clothing with a distinctive logo. You do not want to wear a shirt with a team logo since that logo will either invite conversation from others who may also support that team or for those who hate that team. Not looking weak is another important skill. If you look weak, you will be seen as an easy target. Being discreet about your preparations for an escape is another part of the skills you need to practice. If you are purchasing gear you think you will need, be aware that a large box can attract attention. When you discard those boxes, it is important to remove an address label that will be a clue about your identity.

Another disruptive factor is making noise. Fugitives have been caught because their pockets contained metal tools or other things that made noise when they needed to be quiet. Your keychain should also be silent and not rattle. As you move through an area in which there are other people, do not travel with the same people around you since that improves the chance that they will either remember you or begin a conversation with you. At the same time, try not make eye contact with anybody. Staying on a safer road is also important. If you decide to travel through a dark alley, you will be inviting an attack by predators that may be looking for an easy target. All these steps are ones that you can identify and employ simply through what could be regarded as common sense. They are subtle enough that they would not make you stand out as a person on the run.

Tactical Considerations for Living Off the Grid

Any person who has been forced into a worst-case scenario needs to be familiar with some of the tactics required for survival. This situation might arrive with dramatic suddenness, so advanced preparation is essential. Your goal will be to evolve into what is called a gray man who is not noticed by others.

A first step will be to radically change your facial appearance. If you have a beard, getting rid of it is crucial even though a clean-shaven face is more easily identified by facial recognition technology. The clothes that you wear should be nondescript and possibly purchased at Goodwill or Walmart. Designer jeans and other distinctive, fashionable clothing should be avoided. If possible, you should carry a neck knife which can be worn under your shirt without appearing to be armed. Another accessory is a money belt that will enable you to carry a large amount of cash so you are not dependent on bank cards. The Snake Eater Tactical nylon belt is useful not just to hold up your trousers but also be used to assist you should you have to undertake difficult physical maneuvers. It can even be used as a tourniquet should you suffer a cut.

While moving into the tactical environment you may have to stay in a hotel for a short time. It is advisable that you use a hotel at which you never stayed in the past. When you go out, be sure to carefully place items to determine if someone has been in your room. By taking a digital picture of the arrangement, you can make a precise judgment about whether or not your room has been disturbed. While in a hotel room, you should have your necessary gear tied to you by a string. This is helpful should you find yourself in danger during the night and need to quickly flee. It is very possible that you may not have functioning lights at such a time, so the string makes it possible for you to instantly grab what you need as you leave the darkened room.

Automotive Vulnerabilities

MUCH HAS BEEN WRITTEN ABOUT THE dangers of cell phones because they effectively and consistently spy on users. While first users viewed the cell phone as little more than a device for making phone calls, it quickly evolved into much more. Less attention has been given to the modern automobile. Obviously, its primary use is to enable drivers and passengers to move from one place to another while enjoying a degree of comfort and security. However, like the cell phone the modern automobile provides an abundance of information about its owners.

As noted above, access to an automobile gives the would-be refugee some important advantages. In fact, throughout most of our lives the automobile has been seen as the key to freedom. Every adolescent, upon securing a driver's license and a car feels empowered and liberated. The importance of a car has also been recognized by governments, many of which have set out to limit automobile ownership. Soviet leader Nikita Khrushchev denounced cars as foul smelling armchairs on wheels. Cars not only promoted immorality, he insisted, but were a waste of resources better used by the defense industry.

In the more prosperous Western nations, automobiles were increasingly available to most people. In order to maintain some element of control, governments insisted on a variety of documentation to keep track of both cars and their drivers. The use of surveillance cameras on main highways gave authorities a systematic way of recording which cars were on which roads. Fugitives were often apprehended through this system. More recently, Radio Frequency Identification (RFID) tags are placed in tires. When you buy those tires, they are registered in your name and, as a result, police can more easily locate your car by the tires upon which it travels.

But with modern innovations and an honest concern to locate a vehicle that might have been stolen, devices such as OnStar have enabled authorities to locate those automobiles. The more advanced vehicular telematics systems have gone far beyond the early theft prevention or driver-assistance functions. They help promote our safety, mechanical reliability, and entertainment as well as a host of other modern conveniences. Almost all drivers love features such as forward collision alert, cruise control, parking sensors, and blind-spot monitoring. With these systems, you are never alone as you drive down the road.

Not being alone can also mean that law enforcement officials are riding with you. One of the best illustrations was revealed in 2019 in Clark County, Nevada. Known as Strippergate or Operation G-Sting this was an FBI investigation into bribes being taken by Clark County Commissioners in Las Vegas. The bribes were offered by a lobbyist who represented strip clubs which were trying to convince authorities to eliminate "no touch" ordinances opposed by club owners.

The starting point for this case came when an angry employee of the Texas Auto Center (TAC) in Austin disrupted services for one hundred TAC customers. He accomplished this act of revenge by using stolen login information to access the TAC system. As he remotely triggered the GPS devices car alarms were set off and engine starting systems were disabled. While TAC was providing this service for used cars purchased by individuals with poor credit, nobody is immune to this. The internal computer system of any car or even its audio system can be used as an entry point for a hacker whether he be a rogue individual or an employee of a police organization.

In this case, it was the FBI that took advantage of weaknesses in the vehicle telematics system. The head of the Clark County Commission was a former policeman who decided the best way to avoid FBI eavesdropping was to conduct all sensitive criminal conversations in his car. Even though the FBI did not have a warrant, the OnStar managers agreed to activate the car's microphone and enabled the FBI to make recordings that were the main evidence in the subsequent trial. In spite of debates over the law and proceedings of the Ninth Circuit Court, OnStar continues to collect information for any reason and at any time.

Initially, vehicle digital telematics represented an innocuous innovation that enabled dispatchers to keep track of large fleets of vehicles. With this arrangement, drivers could communicate with central management, record locations, and make projections about arrival times. Eventually it became much more and was seen as a way to improve driving habits through utilization connected car features. Details about weather, tracking, and road conditions were useful in accomplishing this objective. But it evolved into a system that presented its own problems and exhibited numerous vulnerabilities.

Exposure to law enforcement, therefore, is possible because of the existence of OnStar and similar systems. The fugitive wanted by the police should exercise extreme caution if he is using a modern, computerized vehicle. However, because of the activities of hackers, an individual who is not running from the law is vulnerable to private individuals who may be searching for him. An abused spouse or an innocent bystander who happened to witness a major crime can be apprehended if his pursuers employ the services of a skilled hacker.

An additional vulnerability is posed by the increasingly common key fob. With its ability to unlock the car and start the engine before a driver is even near the car, it seems to be an asset for a person fearful of an attack in the parking lot. Short-range radio-frequency (RF) transceivers such as the key fob provide an illusion of safety when, in fact, they are not effective as security devices. The weaknesses of radio-frequency devices have long been obvious in military and intelligence operations. As a civilian application, the RF-powered key fob broadcasts throughout its area of operation and can be easily hacked. It this occurs, the vehicle operator may be locked out of his car or in some other way be the victim of a denial of service attack. The fugitive who is relying on this system to give him an advantage may find himself in the clutches of the pursuers.

The deficiencies of the radio-frequency system are seen in more than the automobile key fob. In an effort to enhance security, modern automobiles use a rolling code system. Unfortunately, the rolling code system uses an algorithm that makes it predictable and easy to evade. This arrangement is effective in providing convenience, but it is not effective if your goal is to ensure safety from intruders determined to steal your car, enter your garage, or kidnap the person trying to escape from pursuers.

While the short-range radio-frequency systems involve serious risks that can easily defeat the fugitive, an even greater and long-term problem is posed by Bluetooth technology. This technical marvel takes data from the user's cell phone, thus giving him easy access to phone numbers and the additional data in the phone. This is, of course, very convenient and the convenience is enhanced by the fact that all of the cell phone information is uploaded into the car's computer database. Thus, not only can modern technology have informed pursuers where you are, by gaining access to cell phone data they can learn about your plans and determine where you may be tomorrow.

The automobile's computer also enables legitimate service providers such as smog detector systems to use the onboard diagnostics ports ("black boxes"). These are located under the dash and can also provide access to those who might not be legitimate service providers. At a minimum these allow insurance companies to evaluate your driving habits but they can also be used in criminal prosecutions if an attorney wants to justify actions against you by examining the motor vehicle event data recorder.

While we may speculate about the absence of systems that will ensure the driver's security or privacy, it is important to realize that these are secondary concerns to manufacturers. Also, if drivers decide they might want to remove these devices, doing so will invalidate the vehicle's warranty. Given the high value placed on personal information in a modern environment, it is not surprising that these systems are useful in collecting data and that much of it has a significant market value. For the fugitive, one of his vital concerns is protection of this privacy and these arrangements undermine his ability to remain off the grid.

Of course, less complex than the ways in which your car will tell people about you is the simple matter of being exposed in a routine traffic stop. They are set up for a multitude of mundane reasons such as checking to see if you are wearing your seat belt. In 2021, Angelo Romero was involved in a routine traffic stop in a small South Dakota town. Romero was an inconsequential lawbreaker who would never have attracted special interest by the local police. A record check revealed that he was facing an outstanding warrant for aggravated assault. He was immediate taken into custody and transported to the county in which the warrant had been served. If he had been on a bus he would never have been apprehended.

Being apprehended purely by chance is a rather common occurrence for fugitives. In 2020, a patrol officer in Ohio noticed a man limping on the side of a road. Being concerned for the welfare of the elderly man, the officer stopped to render assistance and took him to the local hospital. He aroused suspicion because he had no photo identification and was unable to identify himself. Eventually, the officer was able to identify the man as William Jones who was wanted in New York for several charges. He had been living in Ohio under an assumed name for several years. Jones would never have been mentioned on *America's Most Wanted*, but he was still wanted and was caught purely by chance.

An Island Destination

THE SEARCH FOR A DESTINATION IS especially demanding and it is important if one wants to live off the grid that he think out of the box, as people say. An attractive option might be a remote island. There is a considerable amount of information about islands as they attract people for a variety of reasons. Most often the reasons are largely a matter of lifestyle. Conducting such a search does not automatically raise questions about the criminal intentions of the researcher. Island living sounds restful perhaps or even exotic when you contemplate the lifestyle. It's not an accident that places like Key West, Florida, the Pacific Northwest, or Alaska are so popular among people who want a remote, fresh start.

Since few people live on remote islands, this must automatically be considered as an option. Barring the most hysterical desperation, one need not contemplate an island that is both remote and unpopulated. One danger of living in any unpopulated place is that the most minimal indication of life would stand out and generate curiosity if not a search. Moreover, the isolation sought by the fugitive offers both advantages and disadvantages. While his remote location might make him harder to find, it also reduces his prospects for a speedy escape when confronted by pursuers.

However, a remote but populated island is more attractive, especially if you're able to leave the country. A place like Tristan da Cunha in the southern part of the Atlantic Ocean, which has a population of fewer than three hundred people, suggests the attractions for a fugitive. The fact that tourists are a rarity reduces the prospect of any accidental discovery. This is one of the most remote islands on Earth and is accessible only by a seven-day boat trip creating a real challenge for any search team. If pursuers actually got to the island, everyone would notice them and, hopefully, warn or hide a fugitive. On the island, a fugitive would be conspicuous, especially if he is a foreigner, would require special skills for living on the island, and would need to ingratiate himself in the island community. It is not a place where one could be lost in the crowd because there is no crowd. Of course, if he had special skills and could provide a valuable service to the islanders, it would improve the fugitive's prospect. As this is a place noted for fishing, a fugitive could do a lot to provide his own food.

Another frequently cited island refuge is Scotland's Isle of Lewis. It is self-sufficient and less likely to be affected by mainland issues. It is also three hours away from the Scottish mainland so you are able to watch for the infrequent boats to see if anyone suspicious looking has arrived. This is an example of a good shelter for a person who wants to disappear.

As noted earlier in the book, there is a fully off-grid island off the coast of New Zealand that offers advantages for someone seeking to disappear. However, it has an inconsistent telephone network, a limited road system, and residents need an automobile which has four-wheel drive in the rainy season. The relative isolation is disturbed by a six-week period in which vacationers come to the island, so a fugitive would need to take shelter during that time.

Equally remote is Pitcairn Island which is known as the home of the descendants of the sailors who were on the *HMS Bounty* when they mutinied. It is located over more than three thousand miles from New Zealand and has a tiny population of fewer than fifty people. While Pitcairn Island attracted attention in 2004 because of reports about its sexual practices, it is not a tourist destination and if one could actually integrate himself into this tiny community, it could be an attractive refuge. That, however, could be a real challenge.

The fugitive Robert Vesco set out to buy his own island and create it as a country. His idea was that his island "country" could have laws preventing the extradition of any fugitive to another country. His greatest concern, of course, was to prevent himself from being extradited to the United States.

In spite of the appeal of an island, an ill-chosen prospect can be disastrous as a British fugitive learned in 2019. A suspected burglar was attempting to outrun the police while traveling at 75 mph on the wrong side of the road. Realizing that his six-minute chase was coming to a sudden end since his tires had been shredded, he jumped from his car and made his way to an island in the middle of a nature reserve lake. His pursuers soon deployed a small boat and, using night vision cameras, easily located the suspect who was suffering from hypothermia and hiding in the weeds. In reporting the incident, a police spokesman observed that the man had simply marooned himself on the island.

Cyber Developments and Living Off the Grid

There are not enough police resources to successfully track all the people who go missing, whether deliberately or through the actions of criminals who might abduct them. In recent years, civilian groups which will search for a missing person have emerged. Trace Labs is one of the most successful of these organizations. It employs private individuals, many of whom are computer hackers and security professionals, who see a challenge is trying to locate missing people. Trace Labs has created a device it describes as a "Search Party" platform that brings people together to compile open-source information and use it to develop clues as to the direction of a person who is missing.

The internet can be divided into three components: the surface or visible web, the deep web, and the dark web. The so-called dark web, which is a huge collection of websites that hide their IP addresses by using anonymity tools, has become an important source of instruments that are useful in dealing with issues associated with living off the grid. Forged identification cards can be secured on the dark web, some of which are said to be very credible and comparable to the real thing. It also opens up the prospect of using RFID chips to store cell credits and maintain access to money that you may have accumulated in anticipation of escaping from the world of your real life.

The dark web also enables people to get drugs that doctors refuse to prescribe for them. Through sites such as Silk Road marketplace, which was established in 2011 and shut down by the FBI in 2013, buyers were able to purchase a variety of illicit drugs and do so in an environment that was much safer than the typical back alley exchange. In 2014, another marketplace known as AlphaBay was created. Within two years, AlphaBay was ten times as large as Silk Road marketplace ever was. There is speculation that the dark web may be a source for documents that would otherwise be unobtainable. For example, it offers uncensored Wikipedia pages. In this respect, the dark web can be viewed not simply as a mechanism for criminals but also a vehicle to provide information that citizens may need in order to understand political and economic issues. In recent years there has been a growth in official censorship, sometimes by the government, and on other occasions by giant tech companies. The dark web provides a forum for the expression of ideas that are opposed by elites who seek to dominate the political narrative. The dark web also provides services that are important to people seeking to disappear.

The dark web is also a place where one might obtain fake passports and academic certifications, items which would be useful for a person attempting to create a new identity. You can even purchase counterfeit currency on the dark web and perhaps be better able to function while living off the grid. Without a doubt, there are likely technologies which are now being developed but have yet to be made public that will create even more barriers to an effective disappearance. At the same time, the dark web offers mechanisms for a successful disappearance in spite of these technologies. Many countries are already working to combine CCTV coverage, biometrics, smartphone data, and health data to compile information that enable authorities to track almost any person. It's already done in the UK and China. When DNA is folded into databases, disappearance will be impossible once your DNA is matched to any previous record of your DNA. Many people already willingly give up their DNA in order to undertake an ancestry search.

The terms *dark web* and *deep web* are often used interchangeably. It is important to note that these may be similar names but they designate two very different things. The dark web encompasses the world of sites that have been hidden to ensure confidentiality or to protect criminal enterprises such as the recruitment of terrorists or the work of drug dealers or child pornographers. Anonymity is provided by bouncing a signal between multiple servers and using different levels of encryption for each server. Before the terrorist attacks of September 11, 2001, terrorist organizations recruited openly using web sites such as Azzam.com. As a result of anti-terrorist initiatives, such open recruitment was replaced by covert efforts. The dark web now accommodates these criminal undertakings.

By contrast, the deep web contains information that is not linked but needs to be protected. If you log in to a site which requires a password or some other type of authentication, you are on the deep web and there is nothing criminal about this. It is probably something you do every day, like checking your email or entering the patient portal to look at your health records. In doing this, however, you will want to avoid places offering free WiFi because they are less secure than a virtual private network. When a user searches the surface web, he does so using links that bring up certain information and are able to go no further with standard search engines. When you search internally on a website, this will give you access to deep web content such as medical records and legal documents. None of this content can be seen by a standard search engine. If your personal email and social media accounts could be so easily accessed, individuals' interests would be compromised. Exposure of banking information or data stored on scientific or academic sites would be even more costly and would undermine corporate and national security.

By contrast, the dark web may enable a person to engage in criminal activities, especially if they have been forced to go into hiding because they are threatened by individuals or sought by law enforcement. Access to the dark web requires a special search engine such as Excavator or the Onion Router, which was actually created and funded in part by the US government in order to give intelligence agents a mechanism for fully anonymous communications. Web sites on the dark web generally end not with .gov, .edu, or .com but with .onion. A key to anonymity is having an abundance of users that will, by the size of their products, create difficulties for someone attempting to identify which information is valid and created by intelligence organizations. If access is too limited, there will be less information, thus easing the burden of those attempting to assess that information.

In spite of its frightening reputation, the dark web offers content that is perfectly legal as well as criminal content. It might host a forum for the discussion of puzzles or literature as well as a forum for people who are concerned about their loss of privacy or about increasing restrictions on freedom of speech. Although these ideas are not illegal, they may be controversial, especially for a potential government contractor, and open expression of such libertarian notions could result in public ridicule. Thus, for them, the dark web is nothing more than a discreet forum for expression of controversial ideas in a time when such ideas might result in a person being "cancelled" by aggressive social media forces.

But it is the illegal or shocking content for which the dark web is best known. It is an avenue for the purchase of everything from human organs to dangerous military weapons such as rocket launchers. You can purchase a social security card for as little as one hundred dollars if you are creating a fake identity. You can also purchase things like login credentials or banking data for a relatively modest charge. Even more disturbing, the buyer can purchase child pornography or videos of torture being done to people or to animals or can even engage a hitman. There are reputedly instances of victims being auctioned off on the dark web, both adults and children. As we talk about making purchases, it is important to note that the buyer does not use a bank card or a check to make payment. In order to maintain anonymity, payments are generally made using the digital currency Bitcoin.

Created in 2009, Bitcoin is a decentralized digital cryptocurrency. It operates outside the control of banks or governments, thus ensuring complete anonymity. It was released by an anonymous group of people online through open-source software. Bitcoin ensures privacy and security. It is legal in the United States and can be purchased using a credit card or even through Paypal. You can even get Bitcoin at specialized ATMs. Most banks will accept Bitcoin deposits, although there is a certain amount of suspicion and some banks limit the amount of deposits using this cryptocurrency. Otherwise, for the person who seeks to live off the grid, Bitcoin is an important alternative to carrying all your currency on your person. With your private key, the user can employ what is known as the Bitcoin blockchain to authorize transactions. This means that the Bitcoin does not have a physical form but simply an online existence.

Among the most notable services said to be available on the dark web are murderers for hire. While there are sites which offer the services of a hit man to eliminate your enemy or a spouse, many journalists and researchers are skeptical about this. Others are less so.

In 2019, there was a case in San Luis Obispo, California, in which a man attempted to hire a hitman to kill his stepmother. Chris Monteiro, a London-based dark web expert, claimed he had hacked into an illicit website that had a series of posts about murder-for-hire services. It was on this website that Monteiro discovered a plot by a Riverside, California, man whom he said had solicited a killer for his stepmother. Monteiro worked as an information technology systems administrator for an internet security company while also investigating dark web sites that are involved in illicit activities. Monteiro said he penetrated the Besa Mafia website, which is supposedly run by a Romanian who employees Albanian thugs. He claimed to have seen various accounts of requests for killings that had actually taken place. There were often photographs of intended victims.

The investigative news program *48 Hours* ran an episode that examined the story of the Besa Mafia organization. The Romanian who runs this international criminal organization calls himself Yura. The program conducted a worldwide investigation that lasted for six months and offered proof that there were contracts that had been made for murders that had not yet been undertaken at the time the program aired. On his website Yura does not offer direct proof that they have undertaken murders, but suggests that interested parties who go to the internet and search for items such as "murder on the street" can find the evidence of his work. The *48 Hours* producers discovered Yura and his website while investigating the shooting death of Amy Allwine in Minneapolis, a case that had also been investigated by the FBI. Eileen Ormsby, an investigator for *48 Hours* contacted Chris Monteiro, noted above, who has also studied the Besa Mafia. The individual who ordered Amy Allwine's murder paid Besa Mafia $12,000 in Bitcoin. There was a long, complicated investigation in an effort to determine who had contacted Besa Mafia to order the hit. During this time, Yura closed Besa Mafia and created a new site called Cosa Nostra.

In spite of the uncertainty in this case, and uncertainty over whether any murders had actually been committed, there have been other examples. Often there will be a delay in delivering the purchased murders with the organization repeatedly offering excuses for the delays. This has led some analysts to believe most of the murder-for-hire sites are hoaxes in which customers make a down payment but then nothing else happens. Where that occurs, it is reasonable to suggest the entire operation is simply a fraud. This is a criminal act—fraud—but not murder.

According to Eileen Ormsby, many of the murder-for-hire sites are authentic and live up to their boasts of having professional hitmen available throughout the North America and Europe. Sites such as Besa Mafia and Cosa Nostra, with their offer of complete anonymity are the most successful of such operations in history. More than murders for hire are on offer and some sites give customers the opportunity to watch captured terrorists being tortured and eventually murdered. Others offer a pay per view experience of watching real gladiators fight to the death. No matter what you might select, the dark web caters to the widest variety of perversions.

Its reputation, grim though it may be, is not actually earned. Many observers speculate that governments encourage people to view the dark web with fear and dread. Since anonymity is one of the most important services of the dark web, governments have an interest in limiting access it. If one believes that the dark web was actually created by governments, this become a plausible explanation. The more people who explore the dark web the greater chance there is for an erosion of its anonymity. While it is a useful tool, it is not likely to be the province of a criminal underworld bent on selling the services of hit men or of terrorists who use it to communicate with confederates in their violent undertakings.

Conclusion: The Psychological Costs of Living Off the Grid—Is It Worth It?

If you calculate a budget to support a disappearance, it should take into account more than just financial considerations. It is important to recognize the psychological dimensions of living a fake identity and living in constant fear of exposure. Most people experience psychological stress from time to time. It can result from pressure at work if you have to deal with a difficult or unreasonable boss. Psychological stress is associated with significant life changes such as losing a job, experiencing the death of a loved one, a divorce, or any number of personal crises. But these situations are usually episodic, rather than constant.

If, however, you have been forced to flee from your established life and go into hiding, this sort of stress is constant and will have a detrimental impact on your ability to function. While there is good stress—eustress—which can motivate and energize a person, someone on the run will more likely experience bad stress, which causes anxiety and decreased performance levels. The fugitive will suffer from headaches, stomach problems, sleep disturbances, confusion, anxiety, and depression.

Chronic stress can lead to heart disease and a weakened immune system, which will make the fugitive more likely to become sick. This adds to one's difficulties because, as a person living under a false identity, it is more difficult to get adequate health care. One helpful habit to alleviate stress is to engage in regular physical activity. This not only improves your general health, but it is an effective way to reduce stress levels.

Popular literature identifies a number of situations associated with stress. Most of these are situations that will be faced by a person seeking to escape from his old identity. Especially common are moving to a new location, losing contact with family and friends, health and financial problems, and experiencing traumatic events such as those which have forced you to flee. In order to survive, the fugitive must be aware of key indicators such as elevated blood pressure, heart palpitations, fatigue, and insomnia. These could bring about a loss of emotional control, social withdrawal necessitated by your effort to remain hidden, and an inability to solve problems that you face. There are some obvious ways to alleviate these problems. However, the most common methods—leaning on friends and family or consulting a counselor—are simply not available to a person attempting to live off the grid.

There are other factors that would make this sort of life even more difficult. Any person who suffers from health problems will continue to face those health problems. This is especially true for elderly people but also for youngsters. And it is all the more true since the fugitive will not have the support from friends and family who would normally support him during stressful times. Anyone who has a substance abuse problem and might indulge in drugs or alcohol to get through difficulties will be even more vulnerable. Of course, as a fugitive you will have to be socially isolated, and that will increase the vulnerabilities for a person who has taken flight. Living in a remote, rural area can aggravate this vulnerability. Something as simple as not having access to information in your native language can also intensify psychological stress.

An especially brutal version of this psychological phenomenon was experienced during World War II by soldiers who fled into the mountains to elude the German forces. An Australian known as Farleigh James recounted his desperate experiences as a fugitive in a foreign land. He was sent to Greece with the Australian Imperial Forces and was taken prisoner when the Germans invaded in April 1941. A month later he and three other prisoners escaped and spent two years living in Crete as a solitary fugitive because his small group decided they would attract too much attention together. Having none of the resources needed for a successful escape, James endured great physical hardships living in villages among people who spoke no English. While he had no money, he was highly motivated to learn Greek and was able to use his language skills to gather some of what he needed. Most important was a forged identification card which helped him to create a plausible identity.

Forced to live outdoors, James learned the difficulties of having almost no shelter all year round. Food was a severe problem but, even worse, was the rapid deterioration of footwear while traveling the rocky roads of Crete. Having a great deal of solitude and ample time to think, James began to feel depression and a loss of hope. While he was not a religious person, he had expected the local clergy to behave in a Christian manner. What he found was that the priests did little to support the anti-German resistance but enriched themselves through wartime opportunities. Those who did join the resistance condoned the most brutal tactics, including the physical mutilation of dead Germans. If they led guerrilla forces, their sadism and brutality matched that of the worst of the Germans. During his times of reflection, James decided that as a person who had killed others, he was no better than those whose behavior disgusted him.

In the spring of 1943, James was able to flee to North Africa and was eventually returned to Australia. He continued to suffer from the psychological trauma of his wartime experience and found himself abusing his family. His approach to life had become a live-for-today affair without feeling any purpose in his life. For James, his eventual escape from the psychological dimensions of his time in Crete came when he became a Jehovah's Witness.

While the saga of Farleigh James had a satisfactory ending, his success was not due to any professional training he may have received. He was forced to live under brutal conditions and employing his intuitive skills. The South Korean spy Park Chae-seo—known as Black Venus by his handlers—was trained by the Agency for National Security Planning (ANSP), the South Korean intelligence agency. He used various forms of bribery to ingratiate himself with members of the North Korean elite. He also helped enrich members of North Korea's Kim family by selling valuable antiques to rich South Koreans. After his mission was over, Park spoke of the tremendous stress of operating at such a high level among the North Korean official family. He even had a thirty-minute meeting with Kim Jong Un in the Paekhwawon guest house in Pyongyang and concealed a recording device in his person during the session. Operating out of Beijing where he established working relationships with many individuals who constituted North Korea's leadership was doubly stressful.

In 1997, Park was exposed and his North Korean work ended. The ANSP fired Park, who then moved to China. In 2010, with a new security chief running South Korea's intelligence organization, Park was arrested. He was charged with passing classified information to the North although Park insisted this was low-level intelligence designed to convince the North Koreans that he could be trusted. He was convicted and spent six years in solitary confinement. Since his release, his security is bolstered by his threat to release the recording of his meeting with Kim Jong Un if any further move is made against him.

The case of Oleg Gordievsky, a high-ranking career KGB official who became a double agent, provides evidence of the severity of psychological stress and its impact on performance. Known by the codename "Hetman," Gordievsky exposed over two dozen Russian and Eastern European diplomats in London who were spies using their diplomatic cover to work against the West. Going through KGB files in Moscow, he identified opinion formers, politicians, and journalists who were able and willing to assist the Soviets. Some of these were classified as conscious 'agents,' who supplied clandestine information while others were regarded as "confidential contacts." For his part, Gordievsky probably did more than any other person to accelerate the collapse of the Soviet system. He joined the KGB in 1962 and from 1974 to 1985, worked for British intelligence while also fulfilling the duties of his official position as a senior KGB officer who had prestigious assignments in both Copenhagen and London. He is also credited with persuading NATO forces to downgrade the 1982 Able Archer exercise to a command post exercise without President Reagan's personal participation. Gordievsky informed the British that the Soviet leadership actually believed Able Archer was a plan to launch an attack against the USSR and that the Soviets were preparing to undertake a preemptive strike against the West.

In his autobiography *Next Stop Execution*, Gordievsky discussed how he lived a secret life as a double agent who could never tell his wife or any other family member that he was engaged in what amounted to a betrayal of the Soviet system. Every action he took, whether arranging for covert meetings with British agents or acquiring information that he would pass on to MI6, the British foreign intelligence service, he had no person in whom he could confide. He feared that if his family learned of this, his own children might well inform authorities and give testimony against him if he was on trial. Because he had time to himself, he agonized over having to live a lie and practice deception in dealing with friends and family. His primary motive was to work to destroy a system he had grown to hate while asking the British to make arrangements for him to eventually escape along with his wife and their two children.

With the passage of each year, Gordievsky's fear of exposure and likely execution grew. He had difficulties sleeping and his drinking intensified. He began to suffer from claustrophobia. As a result of this tension, in the final stages of his work as a double agent, he began to make mistakes. The first step in triggering his escape plan—Operation Pimlico—was to stand on a certain street corner at an appointed time dressed in a certain manner. His first try was undermined because as a way of explaining his presence at that spot, he bought cigarettes and began to smoke. Unfortunately, the person who was to meet Gordievsky knew of him as a non-smoker so he did not approach him. His next effort to signal the British involved him going to the second floor of St. Basil's Cathedral. Once inside the cathedral, he realized that none of the other men there were wearing hats while he was required to wear a hat. In the heat, this caused excessive sweating which only got worse when he realized the second floor was closed that day. It was not until his third effort that his presence was noted by the British, thus paving the way for his journey to the Finnish border.

His detailed exfiltration instructions were written on a paper which he kept by his bed each night as he readied for his escape. The door to his apartment was barricaded and the sheet of paper was on a napkin on a plate by his bed with matches at hand so he could burn the paper if he heard police breaking into his apartment. As he made his way to the train station where he was supposed to catch a certain train, he arrived several hours too early and had to occupy himself in a way that would not attract attention. Once on the train, he was in fourth class where he had a top bunk. He had to take a double sedative so he could sleep but he fell out of his bunk and injured himself. Because he was bleeding and dirty, he attracted negative attention and was treated rudely by students on the train. To get away from them, he bribed a conductor so he could stand in a compartment at the back where he would not have to face their hostility.

Upon leaving the train at the assigned station, he was to catch a bus that would take him to a small town closer to the Finnish frontier where he was to meet two cars from the British Embassy. Unfortunately, he missed his bus stop and had to feign illness to get the driver to let him off the bus. With this, the number of people who could remember Gordievsky grew ever larger. When he finally arrived at the town, he found a café where he bought food along with an extra beer to take with him while he waited at the meeting point in a nearby forest. After he drank the beer, he threw the bottle away only to realize that it would hold his fingerprints. He was able to locate the bottle. In a similar fashion, he abandoned his map when he realized it was useless, only to have to go back and find it when he remembered it could serve as a clue. As he grew weary of waiting for his contacts, he started to walk down the road thinking he could meet them early only to realize that could throw them off their schedule, especially if they felt KGB agents might be following them. So he walked back to the designated meeting point.

At that point, his British contacts arrived only a few minutes after their scheduled arrival time. Being off the road, they were able to discreetly pack Gordievsky into the trunk of one of their two automobiles. One thing that had worked to their advantage on this particular day was that the area was flooded with young people traveling from Finland to Moscow to participate in an international youth festival. As a result, Gordievsky and the British cars were less conspicuous than otherwise would have been the case.

The British convoy had to cross several checkpoints as they neared the Finnish frontier. Gordievsky was equipped with a thermal blanket which would defeat heat sensors used to determine if a person might be hiding in a car. As dogs began to circle the car in which he was hiding, one the two wives opened a bag of cheese-and-onion chips while the other dropped a dirty diaper under the car to disguise Gordievsky's smell. With that accomplished, the convoy crossed into Finland and met Valerie Pettit, Gordievsky's handler, who had devised Operation Pimlico.

For many people seeking a fresh start, the trouble is worth it. For many, it isn't. One must consider all of the issues raised in this book before undertaking a major life change like going on the run. If one is simply seeking a fresh start, then by all means, give it a try. But if you're on the run from the authorities, life will certainly be difficult. And you should probably get used to looking over your shoulder.

About the Author

John Kiriakou is a former CIA counterterrorism officer, former senior investigator for the Senate Foreign Relations Committee, and former counterterrorism consultant for ABC News. He was responsible for the capture in Pakistan in 2002 of Abu Zubaydah, then believed to be the third-ranking official in al-Qaeda. In 2007, Kiriakou blew the whistle on the CIA's torture program, telling ABC News that the CIA tortured prisoners, that torture was official US government policy, and that the policy had been approved by then-President George W. Bush.

In 2012, Kiriakou was honored with the Joe A. Callaway Award for Civic Courage, an award given to individuals who "advance truth and justice despite the personal risk it creates," and by the inclusion of his portrait in artist Robert Shetterly's series, Americans Who Tell the Truth, which features notable truth-tellers throughout American history. Kiriakou won the PEN Center USA's prestigious First Amendment Award in 2015, the first Blueprint International

Whistleblowing Prize for Bravery and Integrity in the Public Interest in 2016, and the Sam Adams Award for Integrity in Intelligence, also in 2016. A second portrait, by the noted Chinese artist Ai Weiwei, is in the permanent collection of the Smithsonian Institution.

Kiriakou is the author of multiple books on intelligence and the CIA. He lives with his family in Northern Virginia.